Race for a Remedy

The Science and Scientists behind the Next Life-Saving Cancer Medicine

Makhdum Ahmed, MD

Prometheus Books

Essex, Connecticut

Prometheus Books

An imprint of Globe Pequot, the trade division of The Rowman & Littlefield Publishing Group, Inc.
4501 Forbes Blvd., Ste. 200
Lanham, MD 20706
www.rowman.com

Distributed by NATIONAL BOOK NETWORK

British Library Cataloguing in Publication Information Available

Library of Congress Cataloging-in-Publication Data

Names: Ahmed, Makhdum, 1982– author.
Title: Race for a remedy : the science and scientists behind the next life-saving cancer medicine / Makhdum Ahmed, MD.
Description: Lanham, MD : Prometheus Books, [2024] | Includes bibliographical references and index. | Summary: "Exploring basic pharmacological insights, cutting-edge science, and the development of new cancer drug treatments, Race for a Remedy will change the way readers think about medicine"— Provided by publisher.
Identifiers: LCCN 2023052245 (print) | LCCN 2023052246 (ebook) | ISBN 9781633889521 (cloth) | ISBN 9781633889538 (epub)
Subjects: LCSH: Cancer—Immunotherapy—Popular works. | Drug development—Popular works. | Cancer—Treatment—History—Popular works | Cellular therapy—Research—Popular works. | Ahmed, Makhdum, 1982–
Classification: LCC RC271.I45 A36 2024 (print) | LCC RC271.I45 (ebook) | DDC 616.99/4061—dc23/eng/20240123
LC record available at https://lccn.loc.gov/2023052245
LC ebook record available at https://lccn.loc.gov/2023052246

∞™ The paper used in this publication meets the minimum requirements of American National Standard for Information Sciences—Permanence of Paper for Printed Library Materials, ANSI/NISO Z39.48-1992

In memory of Maesha Mahajabeen

Let us go then, you and I,
When the evening is spread out against the sky
Like a patient etherized upon a table;
Let us go, through certain half-deserted streets,
The muttering retreats
Of restless nights in one-night cheap hotels
And sawdust restaurants with oyster-shells:
Streets that follow like a tedious argument
Of insidious intent
To lead you to an overwhelming question . . .
Oh, do not ask, "What is it?"
Let us go and make our visit.

—T. S. Elliot, "The Love Song of J. Alfred Prufrock," 1915

Contents

Note to Readers

The opinions, perspectives, and statements in this book are mine or those of the individuals quoted. None of my current or prior employers have any official position on or make any endorsement of the subject matter of this book, and no public or private company provided any input for this book or offered any endorsement. A lot of the observations are contextual and anecdotal—they should not be considered formal recommendations or suggestions. I took notes for nearly all my personal communications and the interviews conducted for this book. The names of people and the codenames of drugs have been changed in some instances.

Introduction

Drugs are indispensable. It has been humanity's quest since the dawn of civilization, and perhaps even before that, to gain knowledge about drugs, to alleviate pain, and to heal. In this book, I used some of my perspectives and stories to set the stage, to walk you through the promise and the perils of drugs and their development adventures. I have been in love with clinical research for almost twenty years now (I still love it) and enjoy sharing its mechanics with a widening audience, both experts and the merely curious.

As a physician-scientist, I am intrigued by cancer-causing pathways and different ways to interrogate them. During my tenure at the University of Texas MD Anderson Cancer Center, I was part of an amazing team in the Department of Lymphoma and Myeloma at the Cancer Moonshots Program, and I observed very closely the meteoric rise of some of the most successful remedies for blood cancers in recent history. Helping to develop life-saving medicines and taking them to the clinic through powerful collaborations between the pharmaceutical industry and academia is the most exciting part of my job.

In a matter of a few years, starting about 2010, we have witnessed a paradigm shift in the treatment approaches for some of the most aggressive cancers, especially for blood cancers. A growing number of patients who would have been given toxic chemotherapy have started to receive super-specific molecules that target their diseases with amazing accuracy. Or in the case of CAR–T treatments, patients now receive a therapy created from their own cells. Suddenly some patients with limited-to-no treatment options have found hope. Those with merely days left can live longer or even be cured. And for the physicians, researchers, and scientists, it inspired a generation of us to try to achieve more through novel immunotherapy treatments. Cancer research is abuzz over these new, targeted treatments and immunotherapies.

These discoveries blew my mind. I believe there will be great remedies for more of the formerly incurable cancers. Our medical scientists of tomorrow just need to find them.

The drug-development industry is at the frontline of this new era of innovation. The practice-changing scope inspired me to join the pharmaceutical industry, and I am happy that I made the move.

That doesn't mean I've closed my eyes to the problems. I am fully aware, for example, of the heinous and criminal roles that some pharma executives have played over the years in the opioid crisis. I am not oblivious to examples of corporate greed—but then again, never in the history of drug making have we put so much passion, skill, and effort into creating new medicines for cancers. I am humbled by the contributions of our greater community in the journey of drug development, including the key roles that cancer survivors, warriors, and caregivers play. We owe them all these breakthroughs.

The book is loosely divided into three parts. In part I, the focus is on small molecules and especially the remarkable rise of the BTK inhibitor, ibrutinib. I played a minor role in this drug's development journey and was privileged to work with some of the influential players in the BTK inhibitor saga. Part II invites readers to the world of monoclonal antibodies, especially the CD38–targeting antibodies. These are a cornerstone of treatment in multiple myeloma, and the development of anti–CD38 class of drugs runs somewhat parallel to ibrutinib's story. In part III, I highlight what I believe is the finest and the most fascinating modality in cancer treatment: cell therapy. Novel modified antibodies, bispecific or multi-target, also feature in part III.

The scenarios described in this book are a reflection of reality, but the science, the competition, the exchanges, the successes, and the failures are true events. These scenes alternate with quick doses of historical background, scientific narratives, pharmacological concepts, and clinical-development steps. In some instances I've replaced the code names of medicines and names of people with pseudonyms. All other information presented is factual, including approved drug names, their targets, mechanisms of action, and regulatory processes. When appropriate, I provided references to scientific articles or books.

Since I don't want to disappoint any prospective reader, here is a clarification: this book is not about the history of drug development, nor does it cover basic pharmacological concepts in detail. Interested readers are referred to other sources in the chapter notes for understanding the scientific principles of drug discovery.

Finally, in this book I try to emphasize the human beings behind drug development—the men and women who are bright and skilled and who dare to cure.

One day that person could be you—the architect of the next big therapeutic.

Prologue

June 2018
Houston, Texas

On a sweltering day with temperatures soaring above 102 degrees, I found myself lying on a hospital bed, staring at the anesthesiologist, feeling somewhat dizzy.

I was being evaluated for lymphoma—a form of blood cancer. Although I didn't have obvious signs, a persistent rise in my white blood cell counts over a period of several weeks couldn't be explained easily. The less sinister causes of high counts were ruled out. Both I and my primary care doctor agreed this needed to be followed up by an oncologist. A lot of cancer diagnoses were incidental, I knew all too well.

A bone-marrow biopsy was the way to peek at my blood cells factory and to potentially confirm such a diagnosis. My oncologist colleague preferred to do the biopsy under anesthesia due to some preexisting conditions I had. I broke my right ankle a few years before this, and my left knee would be fractured a few months later, which might imply I was a professional football player. But I was no athlete—I just had an uncanny knack for injuring myself in the oddest possible ways (such as fracturing my feet while trying a new footwear—crazy, I know).

Usually, a family member accompanies the person who is about to undergo a painful procedure to potentially confirm such a devastating diagnosis of cancer. I couldn't manage that.

My wife was very pregnant at the time with our son, and she would give birth to him about a month later. Battling her own conditions and complications, she was taking care of our two young daughters at the time. She insisted on going with me—but who would take care of our children? I would not allow her to go, given her own situation. My coworkers respectfully declined to help since the biopsy fell on a weekend, and they would be out of town.

I was in a difficult position. Should I go alone?

Turns out, there are people who provide care to those whose loved ones cannot accompany them. At the MD Anderson Cancer Center, my employer at the time, this wonderful, elderly Houstonian lady agreed to be my caregiver for the day. She played my mother or aunt, pushed the wheelchair around, stood beside me when I blacked out, took me to the recovery room, and out of sheer generosity drove me home to my family afterward.

I have a complicated relationship with chronic pain: it has been my lifelong companion. I expect this uninvited friend to be with me for the rest of my time in this blissful hell, and there is not much I can do about it. Although my pain tolerance is a lot higher than the average individual's and I don't usually complain, an amazing thing happened after the anesthesia—I didn't have any hint of pain anywhere in my body, possibly for the first time in my adult life. This was so unbelievably awesome—I felt like lying there all day long.

"How's it going, doc?" asked the anesthesiologist. "You just passed out!"

"Never been better!" I was jubilant. "Can't really blame Michael Jackson for using that s—t"—that was my nod to Propofol—which was the agent used to knock me out for the procedure.

"Take it easy, my friend," he said with a smile. "You will be confused for a while. Don't marry or divorce anyone in the next few hours."

I accepted the very sound advice and did none of that, although I vaguely remember asking a pretty nurse to go out to lunch with me later, to which she said, "I will write it down, Dr. Ahmed—in case you are confused!"

Then began the scariest period of my life: waiting for the results to come back. What would it show? Do I have lymphoma?

I was no stranger to cancer. I treated cancer patients, and I am a physician-scientist focused on battling cancer for a living. My professional life has been centered around identifying and testing new treatments for aggressive cancers, specifically, cancers of blood cells. And yet, through these moments of anxiety and dread, which started with my visit to the oncologist in my own workplace, I experienced for the first time what thousands of others experience every day in different parts of the world.

How far has it spread? How aggressive was the tumor? What's survival like for different types of lymphoma?

In a matter of days, it all went upside down for me. I didn't have a grip on anything that was happening, as if I were watching a fast-paced movie that barely gave any time for the audience to recover from the action. Certainly, my employment at the nation's premiere cancer center helped, but I began to realize nothing can prepare anyone for such situations. Not even me, who was seasoned by witnessing the impact of the tyrannical emperor of all maladies pretty much every day. This experience brought me closer to my patients and their caregivers than I would have ever imagined.

My bone-marrow aspiration test results came back after about two weeks. I didn't have lymphoma or any other unusual cell. But I did have something: certain suspect mutations or changes in the cells collected from my biopsy.

I reviewed the results with my oncologist.

"What does it mean?" I was curious. It was not nothing.

"They tested your bone marrow aspirates for some commonly known genetic changes or mutations that are associated with cancers," the oncologist, for whom I have the deepest respect, replied.

"Did we ask for them to test it?" I asked. Turns out, the oncologist did not ask for testing for potential cancer-causing mutations in my specimens, but the molecular pathologist did it anyway. They called it panel testing, which meant checking for certain high-risk mutations in patient samples.

I wasn't particularly impressed. What was I supposed to do with this knowledge? Either you found cancer, or you didn't—but this mutation was a cliffhanger.

So I tracked down the molecular pathologist who signed off on the report the following day. We were actually very close—working on a project together at the time.

I showed up at her office with the results in my hand, "What to do with this?"

Apparently, it was a routine procedure to run these genomic analyses in biopsy specimens. In my case, unfortunately, some hits were found. The abnormal genes were circled with a red marker in the test report. These genetic mutations were not actionable, meaning there wasn't much I or the oncologist could do about it other than follow up routinely.

"And needless to say to you," my oncologist stressed, "don't sit around if you feel anything unusual."

What these mutations really revealed was that deep in the DNA of my cells, in the marrow, there were possible roots of abnormal growth. Did it mean I would have cancer soon? Not really. Neither did it mean that I had no such risk. It all had to be correlated clinically. Since I had no abnormal cell identified, a diagnosis was not made. The decision to return for regular follow-ups was a no-brainer.

Thankfully, my high white blood cell counts started to drop in subsequent tests, and I went into a watchful waiting mode. But that didn't leave me with much peace. What if the behavior of my lymphocytes changed drastically? What if they became aggressive? In essence, I became a citizen of "cancer land": a constant state of anxiety for the unknown. It was like a killer lurking in the shadows, waiting to attack me. I became that spy who constantly looks over his shoulder, always looking for his assassin.

I couldn't help thinking about my family. With three young children, my wife would be in a helpless situation if I wasn't around. Several iterations of events continued to play in my mind constantly—What will I do if a diagnosis is reached? What remedy will I seek?

In the world of cancer treatment, chemotherapy is an essential tool. Whether it's blood cancers, such as lymphoma or leukemia or other forms of cancer in different organs (we call them solid tumors), treating with a chemical agent, or a few chemical agents together, is strongly considered. Of course, surgery and radiation are two major treatments.

The principle of using chemo to treat cancer depends on the main character of cancer cells: they grow rapidly. Chemo kills fast-growing cells and mounts a strong defense against the aggression of cancer cells, treating fire with fire. But some normal cells, which also grow fast, such as hairs, nails, and normal blood cells, became the collateral damage of this war between chemo and cancer.

This is where targeted treatment bursts into the cancer therapeutics scene. Instead of attacking all fast-growing cells by a chemical agent (or several agents), these newer treatments look for features that will ideally be present only in cancer cells and not in normal cells. The targeted treatments destroy the cancer cells, preferentially affecting the normal cells minimally.

To create such a targeted treatment, scientists need to figure out the peculiarities of cancer cells—a change, or abnormality not always visible on the outside of the cells but may present inside of them as an unmistakable identity. These peculiarities or cancerous changes can be mutations in genes, like what was found in my samples. But the trouble is, how do you know that the mutated genes are the driver of cancer and if they are, how will you target these with a drug? That requires sophisticated analysis and careful interpretation of genetic data, as well as a thorough understanding of the link between mutated genes and cancer.

If someone had pirated this book you are reading now, before we could declare that copy of the book as pirated, we'd need to have an original title to compare it with. Page for page. Similarly, we needed a complete collection of all genes in humans, also known as the *genome*.

Consider a library full of books as the genome, and each book in that library as an individual gene. Once we have a full library, we can then take out a book and compare that with the pirated copy. A mutated gene needs to be compared with an unmutated gene to understand the differences between them. Then comes the part of creating drugs to target these specific genetic abnormalities, if possible.

A significant challenge for researchers has been that no reference library for the human genome existed—at least not until the start of the new millennium. This changed with the successful completion of the human genome sequencing in the year 2000. What followed was, in my opinion, a golden period for targeted treatments in cancer care—from non-chemo drugs with relatively simpler chemical structures to versatile monoclonal antibodies to immunotherapy, including unique cell therapy.

This is a story of drug development adventures. Our tale will take you to the workshops of pioneering scientists like Paul Ehrlich and to the cutting-edge laboratories of modern-day drugmakers. We will walk the halls of some of the finest hospitals in the world with cancer doctors and researchers who chaperoned revolutionary immunotherapy. Throughout this journey, our patients—survivors and warriors—will inspire you to embrace positivity.

PART I

The Big Journey of a Small Molecule

Chapter 1

Decoding the Map of Life

In science credit goes to the man who convinces the world, not the man to whom the idea first occurs.

—Francis Galton

THE WONDROUS DATABASE

June 26, 2000
East Room of the White House,
Washington, DC

A race was on before the very eyes of the public—a sprint to decode the human genome, the collection of information hidden in the chemical bonds of our DNA. The ever-mysterious unit of biological information—the genes—collectively called the genome, was about to be demystified. At least that's how it was portrayed by newscasters around the world, leading up to the summer of 2000.

President Bill Clinton, flanked by the leaders of the public genome project, Craig Venter and Francis Collins, entered the East Room as I watched a live broadcast of the event in the dining hall of the dormitory I was living in at the time among a group of fellows.

Clinton approached the lectern as Prime Minister Tony Blair of Britain joined the meeting via video call.

The declaration of human genome sequencing didn't register as a breakthrough to me. It wasn't immediately clear how this would lead to new treatments for serious diseases. But the science of genes intrigued me much, courtesy of a high school biology teacher who introduced the concept of heredity and genes. So I decided to pay attention.

"Today, the world is joining us here in the East Room to behold a map of even greater significance." Clinton compared the human genome to an early map of the American frontier created by the courageous first explorers of the continent of North America about two centuries ago—one much admired by President Thomas Jefferson. "We are here to celebrate the completion of the first survey of the entire human genome. Without a doubt, this is the most important, most wondrous map ever produced by humankind."

What kind of map can be called wondrous if it doesn't lead to an amazing treasure? I wondered as my roommate declared it was just a gimmick. Son of a successful surgeon turned politician, my friend was not known to with-hold his opinions—especially when it concerned political actors. But I kind of agreed with his view.

It was hard, even for an eloquent speaker like Clinton, to convey the significance of the project, to explain what it meant for humans. We, the superior biological species on this planet Earth, like to think in a clearly directional way. It's inherently difficult for us to completely understand the idea that every human—every single species on Earth for that matter—carries genetic codes inside their cells that, with the proper tools, can be deciphered. This decoded knowledge can be used to advance science and to possibly cure diseases.

For us, early career medics, codes implied playing with numbers and com-binations, which we typically associated with computers and robots. And yet, we humans hold inside ourselves the greatest codes of life, which determines a lot of things for us even though we may not know it.

"Genomic science will have a real impact on all our lives," Clinton declared. "It will revolutionize the diagnosis, prevention, and treatment of, most, if not all, human diseases."

It sounded completely fascinating but over the top.

"In coming years, doctors increasingly will be able to cure diseases like Alzheimer's, Parkinson's, diabetes, and cancer by attacking their genetic roots. Just to offer one example, patients with some forms of leukemia and breast cancer already are being treated in clinical trials with sophisticated new drugs that precisely target the faulty genes and cancer cells, with little or no risk to healthy cells." Then he quipped, "In fact, it is now conceivable that our children's children will know the term cancer only as a constellation of stars."

This was brilliant from a speech perspective, fitting for the grandiose occa-sion. But the optimism sounded far-fetched to my ears. How could cancer be only known as a constellation soon when it meant a death sentence for the vast majority of us today? When seemingly all that the medical community could do was to delay the inevitable with rounds of chemo when the disease has progressed? I had fresh memories of a patient whom I had started on chemo merely a few days ago. I wished we had something better to give him, something that wouldn't make him so sick, a new medicine that could cure him and would prolong his life meaningfully without causing suffering from the medicine's side effects.

The day's proceedings played out in the White House.

Applause filled the East Room as Tony Blair spoke from across the Atlantic over the video feed. "I would like to single out the Wellcome Trust, without whose vision and foresight, Britain's 30 percent contribution to the overall result would not have been possible. And I would like, too, to mention the imaginative work of Celera and Dr. Craig Venter, who in the best spirit of scientific competition, has helped accelerate today's achievement."

Venter was a central figure in the genome sequencing saga. Born in Salt Lake City, Utah, in 1946, he was a Vietnam veteran. Experiences in the war left Venter wanting to understand more about human mortality and ways to confront it. He became a scientist and was initially drawn to protein molecules, which take care of many of the functions inside cells. His focus moved on to molecular biology and inevitably, to genes. To me, Venter represented the boldness and ambition of groundbreaking research. His grand vision was to categorize all human genes and translate this knowledge to drug target discovery.

On the day's event, Blair's nod to Venter and Celera Genomics, the company Venter founded, was a testament to the role the maverick scientist played in the human genome project. Venter was provocative and often rebellious: a force when it came to molecular biology. A veteran of the National Institute of Health (NIH), Venter had earned a reputation as the bad boy of biological science. Ever so entrepreneurial, he saw opportunity in patenting human genes with a view to restrict genetic information to the privileged. Venter filed preliminary applications to patent some 6,500 human gene sequences in 1999, which drew a lot of fire from academics. What's more, his company Celera Genomics expected to keep filling these patent applications for nearly 20,000 to 30,000 genes—although Venter presumed only about 100 to 300 of these patents might ultimately be kept, a number he'd previously mentioned to the U.S. Congress. The Clinton administration's efforts to combine Celera's genome sequencing endeavor with the NIH's public project was a way to stem the controversy that, after all, "God's handwriting" should not be controlled by a private corporation.

"In a few moments, we'll hear from Celera President, Dr. Craig Venter, who shares in the glory of this day, and deservedly so because of his truly visionary pursuit of innovative strategies to sequence the human genome as rapidly as possible." It was Clinton's turn to shower praise on Venter. "And I thank you, Craig, for what you have done to make this day possible."

It was not accidental that two top leaders on both sides of the Atlantic took a moment to congratulate Venter first. The speed at which the human genome was sequenced was largely attributed to his unorthodox, and sometimes, brash approach to scientific discovery. This public show of celebration and solidarity came after months of politics, arm-twisting, and active engagement between Venter and Collins, aided by the Clinton administration.

Before Venter could take the lectern, the room would hear from Collins.

"Science is a voyage of exploration into the unknown. We are here today to celebrate a milestone along a truly unprecedented voyage, this one into ourselves," said Collins.

At fifty, Collins was a polar-opposite figure to Venter. Taller, leaner, with a flock of boyish hair, Collins was also a firm believer of God—an evangelical. He spent several minutes explaining the importance of the project's success before handing it over to Venter.

"Articulate, provocative, and never complacent, he has ushered in a new way of thinking about biology." That's how Collins introduced Venter to the audience.

With a smile on his face, Venter approached the microphone. Then he stooped behind the lectern, appearing to position a small platform, before stepping on it.

"I am shorter than the previous two speakers," he said. In true Venter fashion, he drew some laughter and applause.

"At 12:30 p.m. today, at a joint press conference with the public genome effort, Celera Genomics will describe the first assembly of the human genetic code from the whole genome shotgun method. Starting only nine months ago, on September 8, 1999, eighteen miles from the White House, a small team of scientists headed by myself, Hamilton Smith, Mark Adams, Gene Myers, and Granger Sutton, began sequencing the DNA of the human genome using a novel method pioneered by essentially the same team five years earlier at the Institute for Genomic Research," Venter said.

He went on to outline that Celera scientists had unlocked complete genetic codes of five individuals, two males and three females, who were Hispanic, Asian, White, or African American.

We now know that any two human beings are 99.9 percent the same in the genome. We are practically identical twins and thus race cannot be defined biologically. But one can appreciate that these racially diverse individuals were selected to perhaps offer a unifying theme that we are not so different after all.

To me, Venter's speech was remarkable for his optimism that the power of genomic data would revolutionize cancer treatments. He said, "Each day approximately two thousand die in America from cancer. Because of the genome efforts that you've heard described by Dr. Collins and me this morning, and the research that will be catalyzed by this information, there's at least the potential to reduce the number of cancer deaths to zero during our lifetimes. The development of new therapeutics will require continued public investment in basic science, and the translations of discoveries into new medicine by the biotech and pharmaceutical industry."

The new treatments that Venter alluded to were targeted treatments for cancer. By sequencing the whole genome, his goal was to uncover specific abnormalities in genes that were potentially risky—that might give rise to cancer or other diseases. In turn, new medicines could target those risky anomalies to switch off cancer. If certain cancer cells have peculiar genetic features, new medicines could target those peculiarities to kill those cells preferentially.

I thought both Clinton and Venter were too bullish about what could be achieved from the genome project. Cancer is a complex disease and often a single cause for cancer cannot be found. And to be clear, targeted treatment was not invented by Venter nor did it spring from human genome sequencing. In fact, the first targeted treatment for breast cancer had been approved in 1977. But when it came to identifying molecular drivers of cancer, the pace of discovery was slow. With genome sequencing, Venter promised to make available the genome that could unlock our DNA, allowing other scientists to work on the genetic basis of disease and possibly accelerating new drug target discovery. He even offered to give the mapped genome in a DVD to any researchers who wanted it, which I thought was artful.

And so, a new millennium kicked off with excitement for creating targeted drugs for cancer. These drugs were of two main categories: small molecules and large molecules. The small molecules were a dominant class in general medicine already. They are molecules that weigh less in terms of molecular weight—under 1,000 Dalton to be precise. From aspirin at the start of the twentieth century to the blockbuster medicines of the present time such as atorvastatin or Omeprazole, the chances of you having several small molecule drugs in your cabinet is very high. These small molecules typically have simpler structure, are taken by mouth, and have predictable behavior once inside the human body. The large molecules are, well, larger and heavier, and they are also known as *biologic*. Although not as ubiquitous as small molecule medicines, biologics were fast becoming a feature in medicine.

For cancer treatment, small molecules were not widely known as effective drugs in the year 2000 with only a few exceptions. We were still very dependent on combination chemo for a wide variety of cancers, and there weren't many small molecules powerful enough to fight cancer cells.

That was about to change.

THE WORKSHOP OF A CHEMOTHERAPIST

1877
Berlin, Germany

Paul Ehrlich, a medical student at the University of Freiburg, peeked through his microscope for the thousandth time. His observations had been confirmed over the past few months on numerous occasions, Ehrlich knew he had discovered something special.

Dyeing clothes to make them colorful had gained a lot of popularity in Europe by the mid-nineteenth century. Royals and fashionistas were in love with their colorful gowns. From Empress Eugene of France to Queen Victoria of England, silk clothes exhibiting beautiful aniline purple became a rage with the ladies. There were competitions on both sides of the Rhine, and several entrepreneurs in Germany and Switzerland were already engaged in a fierce race to dominate the new global market as the demand for dye staining in the clothing industry skyrocketed.

But Ehrlich, a laboratory scientist and a medical doctor, had something else in mind. He applied dyes to different biological cells and tissues. Some cells picked up color quickly while other cells or tissues did it more slowly or not at all, Ehrlich observed.

What happens when I use other types of dyes? Ehrlich asked. He was a perfectionist: in his world, answers could only be found through repeated, precise observations.

He went on to experiment with different dyes and concluded there must be a relationship between cells and dyes. Depending on the attractions between specific dyes and cell types, it was possible to separate them under the microscope. Some dyes mixed easily with water, and these dyes happened to stain most tissues. It was possible to separate different cell parts too—nucleus, cytoplasm, granules—based upon the level of staining.

But he seemed to not be able to stain brain cells or spinal cord cells at all, Ehrlich noted in his lab journal. This resistance to dye uptake by nerve tissue would later come to be known as the *blood-brain barrier*—a boundary that determines what enters the brain.

Ehrlich honed his dye staining techniques in the following years and soon gained a reputation. No one was able to match his talent at devising methods to stain various cells in the human body as well as bacteria. He separated different blood cells—red blood cells, white blood cells, and others—by dye-staining methods, something today's modern hematology laboratories still do. Soon after graduating from Freiburg, Ehrlich went on to work in the Charite Hospital in Berlin along with Robert Koch, a pioneer of microbial causation of disease theory.

More than a century before human genome sequencing, Ehrlich's attempts to identify different cells and cell compartments had a similar goal: the creation of new treatments. By genome sequencing, scientists aimed to identify and target specific molecular features that would only be applicable to a cancer cell, with a view to kill it. Ehrlich's methods of staining and categorizing cells also resulted in a therapeutic approach. He envisioned that when cells become infected with a tiny microbes such as a bacterium, it might be possible to kill the microbe using chemicals without harming the host cells. He called such a chemical agent a "magic bullet"—a form of targeted treatment that would only eliminate the infecting microbe, leaving the host cell unharmed.

In his quest to create a magic bullet for malaria parasites, Ehrlich used a dye named methylene blue. He treated two malaria patients with it successfully, but the benefit of the dye as a treatment wasn't better than the standard treatment for malaria at the time. Ehrlich concluded that although the dye had attractions for the parasite, unfortunately it had toxic effects on normal cells as well.

Ehrlich's search continued for finding a better chemical agent.

Finally at the dawn of the twentieth century, Ehrlich created a new compound to treat another infectious disease that was wreaking havoc in the world: syphilis. The molecule used was a chemically modified form of arsenic, called arsphenamine, created in Ehrlich's laboratory in 1907. It took him a few more years to present the data to the scientific community. And when he did, Ehrlich coined a term to describe the actions of a drug class that became synonymous with cancer for the next century and beyond.

"Now, ladies and gentlemen, I may perhaps take the liberty of inviting you to look into the workshop of the chemo-therapist." These were his words at the 17th International Congress of Medicine in 1913. The arsphenamine molecule—the first chemotherapy—was called 606, that is, the six hundred and sixth molecule tested by Ehrlich and his colleagues in his laboratory against infectious organisms. The molecule came to be known as Salvarsan—the "arsenic that saves."

Soon the toxicity of Salvarsan came to light. It had to be used for a long duration, more than eighteen months, and the patient needed twenty injections to complete full treatment. Moreover, Salvarsan had to be given in combination with other drugs, which resulted in unpleasant side effects such as nausea and vomiting. Ehrlich needed a better molecule with a more specific effect on the disease-causing microbe.

An enhanced version of Ehrlich's molecule, Neosalvarsan, was marketed in 1914, and it shortened the duration of treatment for syphilis. Unfortunately, the side effects persisted. It was not until 1930 that scientists identified the active compound of arsphenamine, which then became the standard treatment for syphilis until penicillin was invented.

Ehrlich did not use chemotherapy to treat cancer patients, although he envisioned a drug class that could preferentially target cells that multiply quickly. Chemotherapy for treating cancer became mainstream during and in the immediate aftermath of World War II. When people were exposed to nitrogen mustard gas, it led to a reduction in their white blood cell count. This early observation prompted further research into chemical agents that killed rapidly dividing cells, and in 1943 a lymphoma patient was treated with a cell-killing chemical called mustine. The next fifty years would go on to establish chemotherapy as the cornerstone treatment for cancer.

But how does chemo work?

Every time a new cell is formed, it must go through a series of phases to become mature, known as the cell cycle. Chemo, broadly, targets the maturing process of the cell either by damaging its DNA directly or interfering with the building blocks of DNA so that the cell can't grow or multiply. The principle of using chemo to treat cancer depends on the main characteristic of cancer cells: they grow rapidly. But some normal cells, which also grow fast, are also destroyed in this war between chemo and cancer. The side effects from chemo are mainly due to this broad-brush killing of rapidly growing cells.

Chemo, along with surgery and radiation treatment, can cure several cancer types, and for some patients with advanced cancer, chemo might be the only option. There is no question that even in modern-day medicine, chemo is a strong weapon against rogue cancer cells that are persistent, aggressive, and rapidly increasing in numbers. As good as chemo regimens are, however, there's room to create more specific treatments, more precise medicines that will target cancer cells only, sparing the normal cells.

Modern-day maverick Venter thought of doing just that: inventing new drug treatments that would change the world of cancer forever through application of knowledge of genomics.

Venter had a plan.

CELERA'S QUEST IN CREATING A NEW DRUG

By the start of the new millennium, Celera Genomics had already made a huge splash on the scientific front: they had succeeded in decoding the human genome and assembling the first draft of the wondrous map. Frustrated by the slow pace of the government-backed Human Genome Project, in 1998 Venter famously launched his own private effort to sequence the genome two years earlier than the public project, which was led by Francis Collins of the National Institutes of Health. In the end it was a tie between Venter and Collins: both reached the finish line together.

In his approach to unlock the genome, Venter employed a previously known technique to identify human genes known as shotgun sequencing. The actual process can be imagined as this: assume the human genome to be a very long thread linked by knots (i.e., chemical bonds). Now, instead of counting all the knots in the original long thread, it is cut into many small pieces. A machine known as a sequencer can read the codes in these tiny pieces of threads. These codes from the random sections of threads are then fed into a computer program, which puts together the long thread based upon readings from the short segments. This method was never used on such a scale as Venter did, and the possibility of errors was there, but Venter envisioned this would save time and money. Scientists could use it as a primary reference to conduct more studies.

Venter knew all too well that the true potential for unlocking genetic information was to find new treatments that had commercial value. He wanted to build on Celera's strength of owning massive amounts of data from genome science and use that information to identify meaningful drug targets. The targets would then be validated and new medicines would be developed. However, all this required expertise and skills in research and development, which Celera didn't have.

And so, Venter introduced his new play, this time in creating drugs. He announced the acquisition of Axys Pharmaceutical Company.

Based in San Francisco, Axys was a thriving biotech that focused on small molecule new drugs that would target key biological processes in cells. Their workforce had strong capabilities in medicinal chemistry and structural biology. With the Axys Pharmaceutical merger, Venter hoped to welcome a skilled team onboard that could form a connection with Celera's genomic data. The goal was to create candidate drugs that would be able to target specific molecular features, exploit any weakness in the disease-causing pathway, and hopefully turn it off.

Celera obtained a few drug candidates from Axys that were still in pre-clinical phase: small molecules that targeted enzymes, which are tiny ribbon-like structures cells employ when they need to get something done. Every enzyme has a special role. They will only interact with very specific molecules to do the task the cell needs. Otherwise, the enzyme remains inert. Tyrosine kinases are one such group of enzymes that control many cell functions. Scientists in Axys had been developing dummy molecules (also known as tool compounds) to bind with tyrosine kinase all in the hope of turning off unhinged chemical processes within the cell.

One of the molecules that Celera scientists created was codenamed CRA-32765. It was just like a thousand other molecules when first created.

CRA-32765 was a Bruton's tyrosine kinase (BTK) inhibitor. BTK is an enzyme generally known for its relationship to inherited diseases. Celera wanted to create BTK inhibitor drugs that would attach to the enzyme in a strong way, thereby blocking the action of BTK, and mainly targeting genetic, immunodeficiency diseases and not cancer.

Celera scientists never thought that their dummy molecule would ever have direct human application. CRA-32765 tended to bind BTK covalently, meaning once this new molecule got hold of the BTK, it wouldn't let go. This covalent or permanent binding was not thought of as a good property for a drug—ideally, scientists prefer non-covalent binding because they want a drug to work on the target for a specific length of time, limiting the chances of side effects. They planned to design a series of similar molecules like CRA-32765 and aimed to compare the strengths of these other molecules with it down the road.

When molecules are created in the lab, they are tested in a group of cells belonging to an organ type. This group of cells, usually collected from patients or obtained from commercial sources, are grown in a lab indefinitely. These are called cell lines, and they can mimic the behavior of the diseases they represent. In other words, these cell lines are models of diseases.

Celera tested the new series of BTK inhibitors in rheumatoid arthritis disease models. CRA-32765 showed activity against rheumatoid arthritis, and it blocked BTK permanently, as expected. The results proved the concept that the covalent molecule could switch off BTK. The question was what was Celera going to do about it?

Nothing, as it turned out.

At Celera's Rockville headquarters where Venter was based, things were not as triumphant as the genome sequencing event in the White House suggested. Venter was a superstar scientist—not a drug developer. His public commitment to turning Celera into a pharmaceutical company wasn't making much headway. Mining the human genome for drug targets, creating new molecules, testing them in disease models, and moving them to the human trials was a complex business, and Axys employees knew it very well, but Venter wasn't having much success in translating genomic data into drug targets quickly, and his grand vision was falling apart. Celera's shareholders and its parent company's top executive wanted Venter to make good on his promise to develop drugs quickly. There was discontent within the company—a stalemate between its star scientist and its business owner.

To be fair to Venter, in drug development, moving beyond experiments in the preclinical setting to creating a real medicine is rarer than one would imagine. Only about 10 percent of candidate drugs make it to approval from their earliest trials. There are thousands of molecules that won't even move to a clinical trial. Even the most successful drug companies go through a game of pick and choose on which project to keep and what molecule to let go, depending on their priorities. Many promising molecules do not pass initial testing, and companies will not talk about them once a project is closed. On the other hand, many successful drugs have been born out of serendipity.

In 2002, frustrated by the absence of a true breakthrough on the therapeutic front, Venter resigned from Celera. After his departure, the company didn't really flourish as a drug creator for the next few years. Because no one at Celera wanted to do anything with the new BTK molecule, CRA-32765, it was never patented and pretty much sat at the bottom of a test tube waiting to be resurrected.

In Sunnyvale, California, another biotech company, Pharmacyclics, had Richard Miller at the helm. Miller was an oncologist and an entrepreneur who knew drug development well, unlike Venter. He had strong connections in the cancer world, had cofounded IDEC Pharma, and had successfully developed a pioneering large molecule drug for cancer, known as rituximab.

Miller received a tip about Celera's intention to sell their molecules. He needed to get hold of a few assets for Pharmacyclics, and the business opportunity sounded appealing. He made a trip to San Francisco to meet with Celera leaders and to learn about their preclinical drug candidates. CRA-32765, still unpatented and neglected, caught his attention.

As a blood cancer expert, Miller knew BTK enzyme was important in promoting B-type cells. If a molecule could stop the enzyme's actions, it could potentially control lymphoma.

Miller worked out a deal.

In 2006, Pharmacyclics bought several molecules from Celera, including CRA-32765, for $6 million. Little did the world know then that CRA-32765, a small molecule, would ultimately transform the lives of many cancer patients and become a blockbuster drug worth billions of dollars.

Chapter 2

The Forerunner

Expectation is the root of all heartache.

—William Shakespeare

THE PROMISE

When Miller acquired the molecules from Celera in 2006, tyrosine kinase enzyme blockers were already a hot topic in cancer, courtesy of another small molecule called imatinib, also known as Gleevec. Created in 1990, and championed by oncologist Brian Druker at the Oregon Health and Science University in Portland, the first clinical trial with Gleevec started in 1998. By spring 2001 it had been approved by the FDA for chronic myeloid leukemia. It was a silver bullet, a beautiful example of how a single-target small molecule can become a life-saving drug. This has hit the world of medicine with a bang.

Miller predicted the ability of tyrosine kinase inhibitors to continue shaking up the field. But when it came to lymphoma—a form of cancer of the blood cells of lymphoid origins, either of B-cell or T-cell types, the role of tyrosine kinases as a drug target wasn't immediately clear.

In our blood, B-cells are the larger lymphocytes that are antibody secreting and play a crucial role in the body's defense against infections: they neutralize viruses and toxins. It is the job of the B-cells to multiply in numbers and rally to defend the body when they detect a foreign agent. This immune defense is usually an orchestrated response with timely actions and retreat.

Occasionally though the signal that rallies the B-cells gets hijacked. It keeps the cells agitated even though no immediate foreign threat is present

15

inside the body. This process can lead to overgrowth, that is, cancers of B-cells like lymphoma.

If a drug could permanently block the signal that was causing B-cells to stay active, could this achieve sufficient control of lymphoma? Miller's thoughts kept coming back to CRA-32765. Through his close connections with Stanford University's researchers, Miller knew of another closely related molecule that appeared to support his hypothesis.

THE OTHER MOLECULE

2005
Palo Alto, California

At Massachusetts General Hospital, Jeff Sharman observed firsthand how cancer was treated with small molecule inhibitors. He saw dramatic results in leukemia patients treated with Gleevec and convinced himself that small molecules could be used to treat other types of blood cancers as well—you just needed to figure out which cancer was addicted to what target.

Sharman returned to California at the end of his residency and decided to work as a hospitalist, taking a break from the intense scheduling of the past few years. From his apartment in Palo Alto, he commuted across the bay to work. But Sharman's goal was to get into the Hematology Fellowship Program at Stanford. He wanted to become a specialist in blood cancers, especially lymphoma.

Lymphomas are of many different subtypes. Some are aggressive while others are slow-growing tumors. Sharman thought the field was advancing rapidly thanks to a better understanding of a series of chemical reactions that were thought to drive lymphomas and were known as *B-cell signaling*.

There was no better place than Stanford for such specialized training—prestige, quality of facilities and great instructors, it had it all. He managed to get into the coveted fellowship program in 2005.

Now at Stanford, Sharman met Sandra Horning, an oncologist and president-elect of the American Society of Clinical Oncology (ASCO) at the time. She was married to Richard Miller.

"She was a lymphoma maven," said Sharman. "I wanted to impress Sandra and learn!"

Horning liked Sharman's enthusiasm. She gave him a project: to do a medical chart review of about thirty patients who had uncontrolled production of lymphocytes after stem-cell transplant. Nothing fancy, but something that gave Sharman an opportunity to prove his skills in clinical research—to show the maven that he meant business.

Sharman was invested.

"I don't think I have ADHD," he said to me in one of our conversations, years later. "But actually, it is possible, at least my wife thinks that way!"

Sharman consulted Ranjana Advani, another lymphoma workhorse at Stanford. Advani suggested he should design a clinical study for a specific nasal type of NK and T-cell lymphoma. These were uncanny tumors with skin lesions and had very limited treatments available. But these lymphomas had one thing in common: they shared a connection with Epstein Barr Virus (EBV). EBV could infect a cell and pretty much hide inside it without being detected. The EBV then hijacks the cell's processes, producing reactions that promote growth of key signaling pathways, leading to several cancers in humans.

Sharman had an obsession for viruses and kinases. He knew that some lymphomas resulted from hyperactive signaling of kinases inside the cells. The role of viruses in promoting kinase actions inside cells was creative, screaming for his attention. Advani's advice on the NK/T-cell project had all three elements that intrigued him: viruses, kinases, and lymphoma!

"I must have studied for three straight days," he said. "I went full manic!"

The fact that EBV hides silently inside a cell, hijacks its functions, and produces certain enzymes that work-up B-cell signaling meant that the lymphoma was dependent on these kinases for growth. What it also meant was that small molecules that block these signaling pathways could be handy in stopping or slowing down lymphoma.

"The EBV was pointing to the B-cell signaling as a crucial element," said Sharman. It wasn't necessarily the virus but the process that the tiny virus kickstarted—the cellular actions orchestrated by the tyrosine kinases—that captivated him.

About two years before this, German immunologist Klaus Rajewsky had beautifully highlighted how EBV promotes lymphomas in mice. Rajewsky showed that once it hijacks a cell, EBV mimics normal growth signals in a B-cell and pretty much tells the cell to stay active, to keep rallying the B-cell lymphocytes—sort of immortalizing them. In fact, in his experiments, Rajewsky illustrated that removal of B-cell signaling in mice led to the animals losing their B-cells altogether. This was a major pointer that tyrosine kinases might hold the key to stopping B-cell lymphoma.

There were three closely linked kinases in the main signaling pathway, and BTK was just one of them. Two other possible candidates, in theory, could be targeted with a drug. Which enzyme of the three candidates in the B-cell signaling pathway was the driver of growth?

Sharman's logic for picking the kinase was simple. "SYK was the first enzyme that kicks off the B-cell signaling cascade, you know," Sharman recalled. "That's why I thought SYK is a good target for a drug in lymphoma."

And so he picked up the phone and called the only company in the Bay area that happened to have a small molecule that targets SYK.

PROOF OF CONCEPT

2007
Palo Alto, California

These were the heydays of biotech.

Companies with expertise in biology and drug designing were enjoying exceptional growth fueled by the genome charting and the promise of new miracle cures. Investors, big and small, came rushing with money. Shares of biotech companies soared while global sales climbed rapidly from around $35 billion in 2002 to nearly $100 billion in 2007.

About a year after Sharman's phone call, a south San Francisco–based biotech, Rigel, identified a safe dose of a small molecule called fostamatinib. It inhibited a specific tyrosine kinase called SYK in arthritis patients. Elliot Grossbard, vice president of development at Rigel, and the Rigel team agreed to start a clinical trial in lymphoma patients with fostamatinib in 2007.

Although the trial was led by the usual heavyweights, Sharman was in the thick of it. He helped enroll sixteen patients in the trial.

Two years earlier, Sharman had finished an elevator pitch to Grossbard with a flourish: "This drug of yours could be a lymphoma medicine."

Grossbard was a trained oncologist. He knew from experience that first-year fellows like Sharman were unvarnished enthusiasts—they don't care much about conventions and pretty much follow their hearts. But Grossbard had once received a similar phone call from another fellow that in the end led to the creation of tissue plasminogen activator (TPA)—a lifesaving drug for heart attacks.

So he listened attentively to Sharman's pitch.

"Listen, you have a good idea but not enough information to start a clinical trial in lymphoma," Grossbard said to Sharman at the end. "But we will share our drug with you."

Armed with a commitment from Rigel, Sharman discussed his SYK idea with Ronald Levy—another of Stanford's brightest minds. Levy had a great sense of new therapeutic discovery. He was the coleader who helped create rituximab along with Richard Miller.

He was highly skeptical, though, about Sharman's idea. "Look, small molecules may as well be bleach. Anything works in vitro, you know. You can kill cells by adding bleach in it—so how is this SYK inhibitor any different?"

Levy's views of in vitro disease models were telling. A lot of cell lines—the disease models—don't represent the disease very well. Cell lines tend to respond to experimental molecules more in a laboratory than in real patients. There are inherent limitations in translating a molecule's action on a lab plate to a living organism.

I have seen the same cell line produce conflicting results in different experiments. It is not at all uncommon to see such discrepancies in cell line data, so a lot of extra caution is needed when interpreting findings of experimental molecules in cell lines. If preclinical cell-line models were always true, we would have cured many cancers by now. I am not suggesting that cell-line findings are always horrible and should be thrown out completely—cell culture models are an important tool in drug development, but their limitations are significant. Every seasoned drug developer knows this.

Despite his reservations, Levy didn't want to totally discourage Sharman. "If you want to come to my lab and work on fostamatinib, that's fine. Show me some preclinical data, and we will see where this goes."

And so Sharman talked himself into working in Levy's lab at Stanford to generate preclinical data with fostamatinib. The idea was to work with lymphoma cell lines and play with different doses of Rigel's fostamatinib to see if the drug worked—typical experiments you would set up to test a new molecule.

He didn't enjoy the lab work.

But Sharman learned a great deal about preclinical research in the laboratory—and promised himself to not get into these kinds of experiments anytime soon. He got what he wanted though: enough data to support a clinical trial with fostamatinib for lymphoma. It was a dream come true for a young hematology doctor to see his ideas leading to a clinical trial during a three-year fellowship.

"The first two patients were quite dramatic," Sharman recalled. "One patient had very large masses under the skin, he used to drive two hours to get to Stanford to receive the treatment. He took the first pill, drove back home, and called me as soon as he arrived. "The tumor is already smaller!" Fascinating stuff—a true moment of joy for a cancer doctor.

Then a second patient, a girl with follicular lymphoma of the lymph nodes, also reported a dramatic shrinking of her tumor.

Clearly, the drug was working.

These were so-called partial responses—meaning the drug was able to reduce tumor size but could not kill it off completely. Still it was promising for a new molecule.

The optimism soon turned to confusion.

Sharman and other lymphoma doctors started to see peculiar findings in patients they treated with Rigel's drug: these patients had a huge number of white blood cells, known as lymphocytes, in their circulation as a response to fostamatinib. The trial started in March 2007 and included eleven investigators in the same number of sites across the United States, including Stanford University Hospital. Typically, the sponsor of a clinical trial hosts regular meetings with the investigators to discuss patient enrolment, new findings, and safety concerns, among other things.

In one such meeting, doctors across the country who had enrolled patients exchanged findings of high lymphocytes counts in their patients. Some even reported observing lymphoma cells in the blood of their patients. That is unusual as you don't expect lymphoma cells in circulating blood unless the patient has overwhelming disease. Usually lymphoma cells are packed in the involved lymph nodes or tissue. In some cases, circulating lymphocytes counts doubled, even tripled within twenty-four hours of the first dose. The fact that they were in the blood meant the SYK inhibitor was presumably causing the lymphoma cells to move from their usual location and be released into the bloodstream of patients—something that is known as *de-margination*. A high white cell count in the blood spells trouble. For example, a sudden surge in a cancer patient could mean an acute crisis—a life threatening event.

"Sometimes, we wondered whether we were doing good or bad to them!" There were anxious moments for Sharman and other doctors who were monitoring these patients. But it was clear from the patients' responses, despite the aggressive nature of their tumors, that SYK inhibition was working.

The fostamatinib trial findings were a bright spot for small molecules in lymphomas. The results were not stellar, but they proved the concept that a tyrosine kinase inhibitor can control lymphoma, producing clinical benefit for patients.

The early data from the sixty-eight patients who were treated in the Phase 1 trial were presented at a plenary session of the American Society of Hematology (ASH) annual conference in 2009. Sharman was only acknowledged as the second author in the presentation and in the paper that followed.

As great as this was for SYK inhibitors and the future of small molecule development addressing lymphoma, I know it was hard for Sharman to accept his name coming second in the paper. I have high regard for his dedication to patients who seek care at community practices and his strength as an advisor in the drug development space. But in academia, these disappointments of not being duly acknowledged are all too common—I have been a witness to and victim of some of these as well.

For Richard Miller, though, the success of fostamatinib was great news: his molecule could be the next star on the block!

Chapter 3

Rebirth

The only place where success comes before work is in the dictionary.

—Vince Lombardi

MILLER'S MISADVENTURE

As rosy as it was for biotech, Miller and Pharmacyclics found themselves in a difficult spot in 2007. If you are a drug-making company with a focus on innovation, you need, well, new drugs. Except the path to drug discovery is complex and costly. And it takes time to develop a pipeline of molecules.

Although Miller secured the deal with Celera and got its six candidate drugs, these were mere promises: hopes that one or more of them might become a money-making molecule. Pharmacyclics required investment and direction for developing these molecules. By then the covalent BTK inhibitor had a new codename: PCI 32765 (the prefix, *PCI*, identified Pharmacyclics as its new owner). Now they needed to move it to preclinical testing before a human trial could be proposed.

But Miller got into an argument with the FDA over its approach to regulation.

This kamikaze urge stemmed from a hefty misadventure Pharmacyclics had had with a drug called Xcytrin. This drug was a key asset for the company and it had full support from Miller and the top executives at Pharmacyclics. But unfortunately for them, Xcytrin was not able to show significant differences in patient outcomes from a statistical standpoint. This means that even though the drug was shown to be beneficial in brain cancer, the benefit of it could have been due to chance alone.

The FDA rejected the Xcytrin application and called it not approvable.

21

This was a strong blow for Miller. His company had spent millions on the development of Xcytrin by then and desperately needed that approval to keep its investors.

Frustrated, Miller took a public stance.

In a *Wall Street Journal* opinion piece on August 1, 2007, he wrote, "The most welcome news a cancer patient can hear from their doctor is 'Your tumor is regressing.' Sadly, the message that the Food & Drug Administration is now delivering to cancer patients is that the fight against tumors is regressing."

Miller went on, "Current FDA policies are discouraging the development of groundbreaking treatments for cancer and other killer diseases, turning the clock back on hard-won regulations put in place in response to the AIDS crisis that allow patients faster access to new drugs."

I never quite understood how locking horns with the FDA publicly could benefit a company. This was scathing and rebellious, and that too coming from a pharma boss whose drugs would ultimately need the FDA's blessing no matter what. But I am no Miller, and he had a reputation as a warrior when it came to defending his ideas and projects. Miller's public feud with the FDA didn't do Pharmacyclics any good. Clearly they needed to pivot, to move on to something bigger and better, to do anything but fight the FDA.

Miller must have thought PCI 32765 could be his way out of this. After all, he was a proponent of including the molecule in the Celera business deal. All his insights and experiences in drug development were telling him a BTK inhibitor might well be a worthwhile option. But even at an age and time when genomic science was advancing our knowledge of human diseases rapidly, using a single small molecule for treating cancer was a bold concept and something not many would bet on.

A key challenge for Pharmacyclics' BTK project team was the lack of preclinical safety data for PCI 32765. A new molecule needs to be tested in cell lines and in animal models (usually rats, dogs, or monkeys) to predict a safe dose of the drug in humans and to understand if it will cause unwanted reactions among other things. The project team had some data from monkeys treated with PCI 32765, but that wasn't enough. They needed a full preclinical package to support their investigational new drug (IND) application to the same FDA that Miller had criticized so harshly. His team also didn't quite know which cancer they should target with it. B-cell cancers are a diverse group with many different types.

Miller came up with a great idea: he went to the University of Colorado Veterinary Program, and they tested the drug in dogs, soon generating data that showed PCI 32765 regressed spontaneous canine lymphomas.

But questions remained about the covalent binding, the tight hug the drug had with BTK. Was it going to cause unwanted side effects or long-term toxicities that would be hard to remedy? Would the risk outweigh the benefits?

Miller had had enough.

"I have patients in the clinic who are dying and need something right away. I can't tell them they'll need to wait around for another year because we have a concern we can't even articulate," he told his project team. This was vintage Miller, the warrior and the dreamer, who would go all in for his ideas.

Of course Miller knew about fostamatinib and the experimental data Sharman had generated at Stanford with it. This was his previous workplace, and Horning, his wife, was actively involved in lymphoma research.

At the time I was doing my postgraduate studies in Australia and attending a community oncology practice in the small town of Wollongong, about sixty miles south of Sydney. I came across the fostamatinib story during a conversation with a professor there.

"You should check the fostamatinib paper, Mak. Cool stuff!" He mentioned it during a lunch break.

These inspiring stories of targeted treatment in action were too few and far between at the time. Still new to the field, I was amazed by the treatment paradigm. The idea that you could have medicines that specifically hit cancer cells and went beyond a systemic, kill-all treatment was so appealing. Understanding the process played out behind B-cell cancer causation is difficult, but this professor's impressive knowledge of molecular biology and his candid commentary on any given tumor behavior was just pure gold. He was a tremendous teacher and a smart oncologist.

I was learning.

I remember reading the fostamatinib trial findings like a novel. Cell circuitry and its biology had intrigued me from very early days, even before I became a physician. The signaling, the controlling of one kinase by another, their ability to influence cell functions downstream was captivating. It sure felt like the complex biological network of cell signaling was starting to reveal its grand purpose: there would be cures for cancer!

THE FIRST TRIAL

2008
California

Sharman had a decision to make.

He had to choose between following his passion for clinical research or becoming a clinician fully focused on patient care. Yes, driving the fostamatinib trial, watching all these patients benefiting from the drug, and realizing his idea of a kinase inhibition proving true was rewarding. But Sharman had also discovered an unpleasant feature of academia: you have to elbow others out of the way sometimes and fight for your own ideas.

I feel the same way having been in an academic environment for many years. Here, being vocal and self-serving goes a long way. If you are not that type, someone who is hungrier or more aggressive will steal the spotlight. Being the introvert that I was, saying, "Stop! That's mine" was more difficult than it should have been. I blame my own limitations and naivety more than anything else for situations like these that I have dealt with.

Then again, people are different.

If that is what academic research was all about, then Sharman felt this was not for him. He was nearing completion of the fellowship and needed to make a financially viable choice. It was at the height of the financial crisis in 2008—he had mouths to feed and bills to pay. A community practice network offered stability and peace of mind over the constant stress of securing grants in an academic setting.

Sharman made up his mind.

During that period, Pharmacyclics already had dog lymphoma data with PCI 32765 in hand and was preparing to start a human clinical trial. At this point, Miller called up Sharman, still at Stanford, and offered him a consultancy contract.

Sharman drove to Sunnyvale, California, Pharmacyclics headquarters half an hour from his home, to advise on the development of the BTK inhibitor. He must have been a great resource to them as there were not many clinicians with knowledge of a new kinase inhibitor that worked on B-cell cancers and especially for a kinase that was closely related to BTK (i.e., SYK), but Sharman had full knowledge of it from preclinical science to clinical experiences with fostamatinib.

I know Miller claims that he wrote the first clinical trial protocol with PCI 32765 (and I am not debating that), but I can't help thinking Sharman had a strong role in the design of the trial given his recent experiences at Stanford. After four visits to Sunnyvale, Sharman left for his new job, and without him, Pharmacyclics started enrolling the first-in-human clinical trial with PCI 32765.

This very first trial included patients with lymphoma as well as patients with another blood cancer known as chronic lymphocytic leukemia (CLL). The same clinical trial wouldn't usually have both lymphoma and leukemia patients, but Miller thought to broaden his reach, wanting to see if the drug worked in cancers that had B-cell features. In the drug development playbook, the first time you start a trial in humans, it is better to have inclusivity—and that means a broad patient population. Once you have a preliminary signal of safety and activity for a new drug, it is then customary to focus on a more-specific patient population.

The Phase 1 clinical trial with PCI 32765 was pretty much run at Stanford, and it took close to a full year to enroll the first cohort of six patients. Taking a full year to fill enrollment for the first cohort usually signals a death sentence for a new drug. It means the drug either causes a lot of toxicity at a low dose or just doesn't generate enough interest from doctors or patients. When you are dealing with aggressive cancers, time is crucial. You just can't have a trial stalled this way.

Miller needed to identify a safe and effective dose for PCI 32765 quickly and then have more patients treated with that dose. He was now at an impasse: either invest more money to keep the trial going or shut down Phase 1.

I try to put myself in Miller's shoes: I have spent tons of money on Xcytrin, which ends up with no approval. I have criticized the nation's health regulators and soured my relationship with them, possibly forever. I believe in a paradigm shift for B-cell lymphoma therapy with a new kinase inhibitor, and I have invested in its first-in-human trial, but the trial has stalled. My company has no money, no other promising drugs, and practically no way out of this hole. And there are no investors eager to come to the rescue because the economy is in a recession.

The writing was on the wall.

In September 2008, Richard Miller resigned from Pharmacyclics, the company he had founded.

FROM THE BRINK

Enter Ahmed Hamdy, the new chief medical officer (CMO) of Pharmacyclics.

Hamdy was a very active and charming man. Trained as a urologist in Egypt, he was working at a biotech company in San Francisco when offered the position of CMO at Pharmacyclics. Hamdy liked what he read in the contract. This would be his first true executive job, and he was hungry for it.

Among the most pressing things that required his attention was speeding up enrollment in the ongoing PCI 32765 trial. Wasting no time, he started to build a team, and among the people he reached out to was Jeff Sharman.

By then Sharman had started his job in Eugene, Oregon, as a cancer expert in the U.S. Oncology Network. Between work and settling in a new place, he had pretty much forgotten about the PCI 32765 trial at Stanford. Hearing about it from Hamdy was refreshing, and Hamdy's offer to enroll patients in the same trial sounded great. Sharman gleefully accepted Hamdy's request to participate in the trial, and between the four community cancer clinics within the Oncology Network, he and his colleagues enrolled a second cohort of patients in the Phase 1 PCI 32765 trial in a single day!

Among these patients were two patients who had chronic lymphocytic leukemia (CLL): Bella and Ana. Both were Sharman's patients.

"I still remember them so clearly, like it was yesterday," says Sharman, becoming nostalgic. "They came on the first day, received PCI 32765, and came back the following day for some blood work. On the second day, Ana's lymphocytes count had tripled."

From his SYK inhibitor experience, Sharman knew this was not essentially a sign of impending doom for the patient but rather of how a kinase inhibitor works.

Sharman could see from his physical exam of Ana that her swollen lymph nodes had actually shrunk. He wanted to wait a few more days ideally before doing anything about it. Upon Hamdy's insistence, he ordered a CT scan, which revealed that the tumor had indeed regressed about 50 percent. Ana felt better and expressed her positive feelings to the medical team.

The second patient, Bella, also had a partial response to the drug.

At long last there was light at the end of the tunnel for Pharmacyclics. The company was very much flailing when Hamdy became its CMO, and now their only drug had started to bring positive news after a long hiatus in trial enrollment.

Hamdy immediately started reaching out to the top experts in the field of leukemia and lymphoma. The reception from the experts was lukewarm: some didn't like the eligibility criteria for the ongoing Phase 1 trial, others were still not convinced by the mechanism of action of the molecule.

Hamdy was paying full attention to all the discussions and to the advice he was receiving. He knew in order to advance PCI 32765 that a well-designed Phase 2 trial was a must. It is okay to start a new molecule with broad inclusion of multiple cancers, but for the molecule to succeed, the team needed to focus on a cancer subtype. All things pointed to CLL as the target—at least from Hamdy's advisory meetings with blood cancer experts—and Hamdy figured there was a conceivable path to approval given some early responses they had witnessed in CLL patients.

So the Phase 2 trial with PCI 32765 was designed to enroll only CLL patients. Hamdy and his de facto project lead, Raquel Izumi, were the queen, bishops, and knights of team BTK. They ran an energetic operation to make sure the trial got its patients.

The early clinical data on PCI 32765 was presented during the ASH annual meeting in Orlando, Florida, December 2010. The Sunnyvale outfit created a lot of buzz in the hematology world because the response rates from patients with CLL were particularly impressive, hovering around 80 percent. But if you were to go by internationally accepted, traditional response criteria, then the actual responders would drop to about 50 percent. The reason for this decrease in response? The sharp rise in the white blood cell counts in patients immediately after receiving PCI 32765. Once that happens, you can no longer consider the patient a partial responder according to the conventional criteria.

In simple words, a partial response means the experimental drug is able to control the patient's tumor progression to a certain degree but not in a complete way. The sharp rise in white cell counts was considered an impending sign of disease progression. Conveniently, Pharmacyclics reported the response rate of PCI 32765 based on the shrinking of the lymph nodes, discounting the massive increase of white blood cells. In this way, they created their own version of the response criteria.

But the company had the support of experienced cancer doctors who were involved in its development and whom Hamdy had consulted extensively.

Susan O'Brien, a major player in PCI 32765 development for CLL, confirmed the uncanny features she observed and what went through her mind when she was treating patients with it at the MD Anderson Cancer Center in Houston. "First of all, people had to realize that their patients were not progressing. Even though the white count was going up, the lymph nodes were shrinking dramatically. So something interesting was going on!"

Let's explain this conundrum a bit more. If we have a patient who receives the molecule and the lymph nodes shrink dramatically to almost 10 percent of their original size, wouldn't we call it at least a partial response? Calling it *stable disease* is not correct. Stable disease pretty much means the patient's disease isn't exploding but at the same time it's not regressing. Clearly with PCI 32765 the patients' tumors were regressing, in some cases quite dramatically.

"We don't want to call it progression and the patients are not really stable because they've actually had dramatic improvement in their nodes," O'Brien commented. "So you may now see the term 'nodal response,' which was developed specifically to try and describe the activity of these drugs."

So now an innovative term *nodal response* was used to describe the molecule's action, and this term soon started to appear in scientific articles.

With drugs like PCI 32765 that targets BTK, we expected to see something entirely different from what would be seen with chemotherapy. While chemo kills circulating cells (that includes white cells), this new molecule caused the white cells counts to go up. That's because PCI 32765 interferes with cells attaching to their supporting tissue, called *stroma* (or *niche*), and the cells are released from that stroma to go out into the circulation—at least that was the theory behind the high white cell phenomenon.

This sudden surge in white blood cells was called a *compartment shift*. The intention behind the naming was to explain the movement of the cancer cells from one compartment in the body (i.e., tissue) to another compartment, which was circulating blood.

When I first read about compartment shift, I envisioned it as spraying the moss off the roof of my house with a high-pressure water hose. If the moss on the roof were cancer cells and the water sprayed from the hose were the drug, then I imagined that the drug just scraped off the tumor cells from their original location and washed them into a draining pool of water on the ground. The pool, in this analogy, is blood circulating in the patient's body. This was thought of as a mode of action for kinase inhibitor drugs because imatinib (Gleevec) and fostamatinib both had similar features. Once the drugs drove the cancer cells away from their home ground and had separated them from their familiar surroundings, the cells were more likely to die off.

A year later, O'Brien presented updated findings from the Phase 2 study with PCI 32765, reporting a nearly 70 percent overall response rate among CLL patients after two cohorts of patients were treated and monitored for up to twelve months. These patients' diseases were highly resistant to other treatments available to them, and their bone marrow was pretty much packed with CLL cells, which meant these patients couldn't have received chemo. Usually patients require a certain level of bone-marrow function because the production of different blood cells needs to be at least in a safe range for the patient to participate in a clinical trial. If not for PCI 32765, they wouldn't be eligible for other trials. They did not have any standard treatments left to choose from.

These were exciting developments for Hamdy, but Pharmacyclics was not out of the woods yet. It needed to secure FDA approval for PCI 32765. Hamdy knew he needed to act quickly, and a plan materialized in his head.

STEP ON THE GAS

When Miller fiercely criticized the FDA over their apparent rigidity to new drug approvals before his resignation as Pharmacyclics' CEO, he noted that the agency was not following "hard-won regulation" created as a "response to the HIV crisis." He was alluding to the FDA's adoption of a program in 1992 for allowing patients access to new medicines and especially those addressing the HIV/AIDS epidemic.

If a disease was serious, there were no available medicines for it, and a new drug happened to offer a possible benefit for the patients, then the FDA could conditionally approve that drug on an expedited basis. The condition of approval could then be lifted later with more data from a larger study. This came to be known as the FDA's Accelerated Approval (AA) Program—which is still in effect.

As a drug developer, my first approach is to look for any opportunity for accelerated approval, especially if I'm working with an aggressive and currently incurable cancer. This does two things: it allows the patient to receive a promising though experimental medicine that might possibly be their last hope, and it allows a drug developer such as me to continue conducting larger trials with more patients to prove that the drug indeed works with more data especially on patient survival. If early responses from a new drug are not validated in a larger trial, the FDA will not grant full approval of the new medicine. So a confirmatory trial may not prove the efficacy of a conditionally granted, expeditiously approved drug, and the drug's approval can be withdrawn. I am aware of at least twenty-six drugs for various cancers since 2011 that received accelerated approval but were subsequently withdrawn from the market being no longer FDA–approved.

Seeing the early response data in preliminary trials in CLL, Hamdy's eyes must have lit up. Could he focus on a blood cancer that was then incurable and aggressive and for which there was no standard treatment available? That would be the key to an accelerated approval!

Mantle cell lymphoma (MCL) was one such blood cancer. This was a B-cell variety as well. Once MCL patients relapsed from chemotherapy, the usual first-line treatment, there were not many treatments available. Doctors around the world would try their own combinations of approved medicines or a last-ditch effort of chemo to try to save patients with relapsed MCL. A new drug taken orally for these patients would be great, and if that drug happened to sufficiently control their cancer, it would be nothing short of extraordinary.

Hamdy was vying for an accelerated approval in MCL, he could see a clear path for PCI 32765 in MCL, but this would prove bittersweet for him.

The formula goes like this: design a good molecule (or obtain one from another biotech through a business deal) and then target a disease where there is an unmet need, which typically means there's no approved medicine at this stage of the disease. Then go for a trial to show your drug helps patients through a surrogate (i.e., proxy) measure of long-term benefit such as months without progression of cancer, known as *progression-free survival* (PFS), with limited side effects, and voila! You have an approved medicine!

Except this is never as easy as it sounds.

Hamdy was a man on a mission. He wanted to engage with doctors who treated MCL regularly, who were hungry, and who were young. The rarer the cancer, the harder it is to find a sizable patient pool who gets treated for it at a hospital where the physicians might be willing to enroll their patients with an unproven medicine.

Hamdy's search for a rare-cancer cure led him to two of the finest doctors in cancer care and an outcome he didn't anticipate.

Chapter 4

Hailing from Qufu

Wherever you go, go with all your heart.

—Confucius

MR. "BTK"

2009
Barcelona, Spain

Michael Wang hurried to the conference hall. His flight was late, and he needed to check in for a presentation later in the day. For a cancer doctor, Wang had a boyish demeanor: he projected enthusiasm, a good amount of it, and he had a strong passion for his patients. In return, his patients absolutely loved him. I was impressed by his energy and his strong dedication to care in the years we have worked together. The man's passion is contagious.

In 2009 Wang was starting to find his place in the oncology community. At forty-seven he was an assistant professor at the Department of Lymphoma and Myeloma at the MD Anderson Cancer Center. I noticed his youthful enthusiasm on our very first encounter there. As a young oncologist, Wang didn't just want to maintain the status quo: he had a real sense of urgency about him, and you could feel his passion for bringing new treatment options to the clinic.

At the time, we didn't have a lot of options for MCL other than up-front chemotherapy. Targeted treatment was a desired concept but with not much to choose from really.

"There is no real Chinese word for *mantle*, you know," I recall Wang saying with a chuckle on our first professional interaction in Houston, on the sixth floor of the faculty center at MD Anderson. He made a shape with his two hands, to model the roof of a house, indicating an elevated place. "I always struggle with my Chinese colleagues to explain the mantle cell lymphoma."

In Barcelona, Wang was heading to the conference hall to present findings from a two-drug combo in lymphoma. These two drugs, known as R2 (Rituximab and Revlimid), were gaining popularity among physicians who treated lymphoma, for they offered a chemo-free option. Still, in newly diagnosed patients chemo was the standard. Some practitioners preferred a bendamustine (B)–rituximab (R) combination to R2.

Wang grew up in the small town of Qufu in the Shandong province of East China, best known as the hometown of Confucius. "I always tell people, I am from Confucius's town," he told me proudly. "Only about 200,000 people live there."

That may be a big number for many, 200,000, but when you consider the population of China, it is really a tiny fraction. Every child in Qufu was taught Confucius's teachings. They were expected to emulate these principles in their lives, and for Wang, he kept Confucianism close to heart.

"If you take care of yourself, your family, and then your community, you will do well," Wang said with belief. "Let's acknowledge that we have a responsibility to our surroundings." But you need to be true to yourself. One must not embellish reality just so it goes with people's perceptions. Wang thought there was a strong tendency to associate poverty with people of Asian origin, and especially when they are successful professionally.

"I was not dirt-poor," he told me in one of our conversations. "I must acknowledge that I had more opportunities than many others who didn't live near the city. I learned English as a kid from a West-educated teacher."

When he started his clinical career in Houston, Wang used to see patients with different types of blood cancers but primarily lymphoma patients. As was the case with other types of tumors, aggressive lymphoma would generally be treated with chemotherapy, which would be two to four drugs together.

Lymphomas are cancers of the blood cells of lymphoid origins—either of B-cell or T-cell types. Less-common cell types include natural killer (NK) cell lymphoma. There are hundreds of variations of lymphomas, but how the disease will behave, and how aggressively it will spread, depends on its cell type. MCL is a rare cancer of the B-cell type. B-cells are the larger of lymphocytes that are antibody-secreting and play a crucial role in the body's defense against infections: they neutralize viruses and toxins. MCL cells develop in the mantle (or ceiling) zone of a lymphatic nodule, which lymphocytes typically inhabit, before moving into the center of the nodule.

It was in Barcelona that Wang first met Hamdy, the newly appointed CMO at Pharmacyclics, which owned PCI 32765 that had now been named ibrutinib.

Wang's colleague Nathan Fowler, an early adopter of the R2 regimen, approached him in Barcelona. Fowler, fifty-one, was a veteran of the U.S. Armed Forces who was also an oncologist at MD Anderson, at that time overseeing early clinical trials in the department. After Wang's trial presentation, he brought Hamdy to talk to Wang about the new molecule. The first impression Wang had of Hamdy was of a very tall guy who talked about things Wang had little idea of.

This new drug in question had the codename PCI 32765. Hearing from Hamdy about the molecule and how it targeted BTK, Wang's first response was "What is BTK?"

"It's called Bruton's tyrosine kinase," Hamdy replied.

"Who is Bruton?"

The question led to much laughter. Wang had his way of dealing with conversations. In fact, BTK was named after American pediatrician and immunologist Ogden Bruton, who discovered a rare condition in children resulting from immune deficiency related to the BTK gene.

Hamdy explained to Wang the status of PCI 32765 development. His vision of getting an accelerated approval in mantle cell lymphoma intrigued Wang.

"I am interested—let's talk about a clinical trial in MCL," Wang said, requesting an update on the drug once more data became available. Hamdy vowed to reconnect.

Returning from Barcelona, Wang found himself completely inundated by work. At the back of his mind, his thoughts kept coming back to PCI 32765. Whatever happened to the tall guy, Wang wondered, didn't he say he would touch base?

Finally, Hamdy called.

Summer 2009
Boca Raton, Florida

The early findings from the first-ever human clinical trial with ibrutinib (PCI 32765) didn't appear to translate into enthusiasm for doctors who treat MCL, possibly because it was the early days for the molecule and the Phase 1 trial wasn't really focused on MCL.

On Hamdy's invitation for a face-to-face meeting of clinicians in Florida later in 2009, only two lymphoma doctors showed up: Michael Wang, and Peter Martin of Weill Cornell Medicine.

The discussions that followed inevitably centered on starting a key clinical trial in MCL patients only, something known as a *pivotal phase trial*. The scope was big, albeit risky, since the patients would be enrolled in the trial after they had failed at least three prior treatments—or three prior *lines*, as we call it. These were patients who had no definite standard treatment left for them, meaning we didn't have any medicines or combinations of medicines that were approved to treat their cancer at the time, though doctors around the world might go for a variety of different options: another round of chemo or other drugs depending on the patient's condition and age.

It's the dream of every cancer doctor to have a chemo-free drug. Patients love targeted treatment because it saves them from unpleasant reactions, long hospital stays, and endless worry. But the caveat is that chemo-free may not cure. You want to give a drug to your patients that is most likely to cure their cancer. With chemo-free treatments, you are always thinking: Should I instead give chemo? Am I taking a chance?

But then again, if there's a chemo-free, targeted drug that has strong data— early proof that it can induce remission—you want to give it to your patient earlier in their disease. That's when the drug has a better shot at beating cancer compared to later when the patient has already relapsed from a few treatment lines, and the cancer cells had more chances to mutate. One in four MCL patients will not go to a second line of treatment, so it might well be too late to save them with a new drug by the time they relapse from the first round of chemotherapy.

"Strike the cancer hard the first time is your best option," said Wang. "I try to give the targeted drug as early as possible—that's how I see it working best." This is where Wang stood out from a lot of the others. He was bold. If he believed a drug could work and had strong evidence from early trials, Wang advocated for its use early in the treatment. Not many would go in this direction. And many would challenge such a position.

At the investigator meeting in Florida, Wang agreed to be the principal investigator for the MCL pivotal phase clinical trial with ibrutinib.

CRABS BY THE OCEAN

David sat in his car at a traffic light near Hillsboro, Florida. Heading north, he was driving to downtown Boca Raton to meet a close friend who also happened to be his doctor, Michael Wang.

David didn't feel very well. For the past few days he'd had bouts of nausea and a choking sensation that seemed to not go away. He ignored the feelings, pretended they weren't there. He'd figured a long time ago the earlier you give in to your symptoms, the harder it is to tackle them. Perhaps it was his defense mechanism against a painful reality: he had a rare cancer in his blood that kept coming back.

Fifteen minutes later, David pulled into a parking lot of a Boca Raton establishment.

"I am here," he texted Wang.

Wang was about to wrap up his day at the investigator meeting hosted by Hamdy—not a super-exciting event, but he was genuinely encouraged by his lead investigator role in the ibrutinib MCL trial. David was one of his closest patients, and he had flown between Houston and Boca Raton to see Wang for the treatment of his MCL over the past few years. Wang thought David's disease was in remission—at least, the scans from his last visit were good. From the moment David learned that Wang would be in Boca Raton, on his home turf, he was insistent they meet: "Let's have crabs by the ocean, Doctor!"

Wang walked over to David's car. "How are you, David?"

"Good to see you, Dr. Wang!"

David really admired the man. To him Wang was someone who brightened his day every time he saw the guy.

They drove to a seafood place by the water.

At eighty, David had quite an appetite. He ordered colossal crabs and a jumbo shrimp combo. From the dining spot they could see the intercoastal waterways leading into Lake Boca Raton just about half a mile from the open waters of the Atlantic.

Sipping his cocktail, Wang felt relaxed. He was looking forward to a night of calm for something told him work was going to get pretty hectic in the weeks ahead.

Sensing something odd, he asked David, "How are you, my friend, is everything okay?" For a moment, it seemed like David was going to reply all was fine in his usual cheerful voice. Instead, hesitating, he replied, "Actually, Dr. Wang, I would like you to see this." He asked Wang to feel his neck area.

David's lymph nodes were so swollen it surprised Wang. They felt like large grapes.

"Goodness, David," he said. "These are very big. When did it happen?"

David hesitated. "Can't tell for sure, Dr. Wang. But I am not feeling well. I feel like I have a mass in my tummy too. Do you think it came back?"

Clearly he didn't want to say it out loud, but deep inside, David knew all too well his MCL had progressed. Yet again.

It was Wang's turn to be the doctor. He understood immediately that David's MCL was in progression and something needed to be done. Wang didn't think David would tolerate salvage chemotherapy at his age and in his physical condition—or at least not without adverse reactions. Of the available options, none seem to be appropriate for him as the possible side effects outweighed the benefits.

Then it occurred to him that David could receive ibrutinib through the clinical trial he had just discussed at the investigator meeting with Hamdy.

"David, I will have a clinical trial opening at MD Anderson soon," said Wang. "It's a new drug—experimental—called PCI now."

"Given with chemo?" David asked.

"No, that's the beauty of it. It's in capsule form. You take it by mouth once or maybe twice daily. It hits the BTK target very strongly and could stop your disease."

David was immediately interested. It gave him a new hope—especially as he had made up his mind against chemotherapy already.

"I trust you, Dr. Wang. Let me know how I can sign up for the trial."

Wang promised to keep in touch.

Two months later, David was among the first patients to be treated with ibrutinib in the MCL pivotal phase trial under the supervision of Michael Wang. The trial marked the start of ibrutinib's journey as a single drug in MCL patients who relapsed from earlier treatments.

SIMON SAYS

Four thousand miles across the Atlantic from Boca Raton, Simon Rule received a call from Ahmed Hamdy at his workplace in the Plymouth Hospital, UK, after morning rounds.

Rule was a major candidate of Hamdy's to lead pivotal trials in Europe given that he was involved in MCL clinical research and saw patients.

"Simon, you are into mantle cell lymphoma, right?" asked Hamdy.

"Yes."

As with the SYK and BTK drugs, Rule had seen dramatic responses to an older drug called thalidomide in two of his patients. Their large lymph nodes regressed pretty much overnight. But because thalidomide had a tragic history (the drug was known to cause serious birth defects), no one would go for a clinical trial with it. So in his attempt to repurpose a closely-related drug, lenalidomide (a.k.a. Revlimid, one of the *R*s in the R2 regimen), Rule wrote to Celgene to start a trial with lenalidomide for MCL at Plymouth Hospital. The answer was a resounding no. "We don't use this drug in lymphoid cancers," said the drugmaker.

With a little tweaking of the dose and schedule of thalidomide, Rule did his own little clinical research with his patients. His sole goal was to offer something new, something that alleviated their MCL progression symptoms.

But he wanted more.

"Would you mind meeting me—I want to show you some data with a new molecule."

"Sure."

Hamdy proposed to meet Rule in London the next Thursday. Taking a day out of his annual leave, Rule got on a train to London. He was genuinely intrigued. As one of the top five hospitals in the country, Plymouth Peninsula served a large population. Yet there was not a lot happening on the clinical trial front and especially for aggressive blood cancers. Rule wanted to not only study new medicines but also repurpose drugs that were already available for other diseases or conditions. In his mind, it was fine to follow new science but the trouble with that was you might miss a clue that was right in front of you. Oftentimes all you needed was just a stroke of luck.

Rule acknowledges the nuances in life. Years ago, his high school headmaster told him that he didn't take life seriously so he should not go to medical school. Yet Rule became the first doctor in his family. At his medical school interview he was asked, "Are you well organized?" Rule replied, "No."

"Then how do you think you will manage the medical profession?"

"I will get myself a decent secretary," was Rule's response, which apparently got him into Nottingham Medical School. Ever so candid, he cares for his patients deeply and looks for every opportunity to do things a bit differently.

In a cafe in London, Rule met Hamdy and Gwen Fyfe—a veteran drug developer who had joined the Pharmacyclics board after a long, successful tenure at Genentech. Hamdy showed Rule early data from the Phase 1 trial with ibrutinib with about fifteen MCL patients (these were the first-in-human trial patients being enrolled by Sharman and others). The data hadn't been published or presented anywhere at the time, and Rule didn't even sign a confidentiality agreement with Pharmacyclics. On Hamdy's laptop screen, one thing caught his attention: MCL patients didn't suffer a lot of side effects from the drug. Chemo typically comes with a lot of unwanted toxicity, but the BTK drug looked clean.

"You have an impressive toxicity profile. I will be more interested to know about the drug's efficacy," Rule said to the Pharmacyclics duo.

"Well, I am off to see Martin Dreyling to talk about his interest in the ibrutinib trial," said Hamdy. "Let's catch up afterwards." Dreyling was a professor and head of the Lymphoma Unit at the University Hospital in Munich, Germany.

In the end, Rule led the UK investigations of ibrutinib in MCL while Michael Wang was the global lead. Rule saw the patients he enrolled personally. Whether it was Phase 1 or 2, Rule would go to see every patient starting with the trial drugs, observing how they did during the drug administration and afterward and sometimes multiple times a day—something he prided himself upon.

The very first patient Rule treated with ibrutinib ended up having huge lymphocytosis (a high number of white cells) on day 8: a count of more than 100,000, which Rule had never seen in an MCL patient before.

Panicked, Rule reached out to Hamdy in Sunnyvale.

The patient was doing well, the lymph nodes had shrunk considerably, and everything else appeared okay except for this crazy white cell count!

"We've seen this happen," Hamdy assured Rule. "The counts will gradually come down."

"Okay."

Rule decided to stay calm and make no changes.

Seen through a microscope, these cells would have scared even the most seasoned professional: leukemic cells floating in their thousands in the bloodstream. Rule recalled that at least two patients in the UK were admitted to emergency departments after their doctors observed these cells because the doctors thought these patients were in a crisis and needed immediate medical intervention. He initially thought of this as a compartment shift event as seen in patients with CLL by Sharman and O'Brien, but later Rule had a different view.

It turned out that the patients who had bone-marrow involvement, meaning those whose MCL cells were in their bone marrow, had significant increases in white cell counts after getting ibrutinib, but those without the marrow involvement did not. This became apparent in the first thirteen patients whom Rule treated in the UK. Later, he published an article in the *British Journal of Hematology* reporting this finding with a much larger global patient population. For MCL at least, the shrinking of lymph nodes wasn't clearly associated with a high white cell count. It was not a case of de-margination.

Chapter 5

A Marathon Worth Running

Cancer is a word, not a sentence.

—Robert Buckman

EARLY WIN

In drug creation, a win can mean different things to different people. For Craig Venter, decoding the human genome and identifying new drug targets that could eliminate deadly cancers was the goal—a spectacular win if that were to happen. For Hamdy, getting ibrutinib to an accelerated approval in a cancer such as MCL would be a fine win and even better if that win would lead to bigger opportunities for his drug.

For Sharman, Wang, Rule, and a whole lot of us, we longed for a non-chemo drug—a pill that would improve our patients' survival and offer them a long remission from their cancer. We would love to cure every single patient if we could, but the path to cure isn't easy. Even if ibrutinib could show benefits in patients who had received multiple other treatments, for the drug to be a major success it had to show benefits earlier in the patient's treatment. In drug designing and in the healthcare industry in general, we love to throw around words that on the surface look simple: take the *benefit*, or *efficacy*, of a drug, for example.

Efficacy is the ability of a drug to produce beneficial changes under controlled or standard conditions. Using this definition, you can replace the word *drug* with any other therapeutic such as *surgery*, *medical device*, or another treatment, but the idea is the same: a beneficial change is happening because of the said intervention. Note that this beneficial change is being produced in a controlled, or standard, condition.

This simply means that the drug (or the intervention) is given in an ideal setting—one where everything is being closely monitored, and that means it's usually in a super-specialized center such as a hospital or clinic, and in a trial. The first pivotal phase ibrutinib trial that Pharmacyclics was conducting for MCL was not a randomized trial. The first trial is usually a *single-arm trial*—meaning patients with MCL who met the criteria were given ibrutinib as a single drug and were monitored periodically to see if their cancer was responding to the treatment. There are no comparison patient groups in an early trial. When a new drug is going through early experimental work, you want to make sure that it does not harm anyone—that's the priority at that stage. At this early phase, the drug needs to be given to patients in a carefully monitored environment so that any unwanted or serious side effects from the drug can be managed by trained medical professionals quickly. Much about the drug is yet to be known in an early trial.

But ultimately the gold standard trial, the one that provides the evidence that determines whether a drug will be approved or not—is known as a *randomized controlled trial* (RCT). In an RCT, we have very specific rules that determine who is included and who isn't (it should be noted this is an artificial scenario, a distortion of reality). And in an RCT, an experimental drug must be compared with an approved drug or drugs. The experimental medicine must show benefit over the approved drug. That's the hurdle each drug must pass to prove its efficacy on the way to its own approval.

While ibrutinib wasn't there yet, there was much excitement and some reason to celebrate in 2011 when data from the first MCL trial became available. One person who didn't join the celebration, or couldn't, was Hamdy. He had been fired from his position as CMO of Pharmacyclics in the same year.

In an unexpected twist, Hamdy's focus on MCL as an accelerated opportunity was not viewed favorably by his employer. Pharmacyclics had brought in veteran players in the development of ibrutinib, players like Gwen Fyfe, who had a big pharma track record. These newer players didn't agree with Hamdy's charted path and thought of MCL as a distraction from the original intent of the drug: approval for use in the treatment of CLL. From a commercial point of view, of course, CLL had a larger patient population compared to MCL (roughly about three times more patients with CLL) and thus offered a better business opportunity for Pharmacyclics. Hamdy's expertise was looked upon as more suited to early drug development (such as Phase 1), but clearly ibrutinib was aiming for a longer ride with more late development (such as Phase 3 or a pivotal Phase 2).

When I learned of Hamdy's exit from Pharmacyclics, I immediately thought of Craig Venter and Richard Miller. Both were pretty much made to resign from their position by their executive boss (in Miller's case, the same boss as Hamdy's), and all of them were related to the BTK inhibitor. In the case of Venter, he never really developed ibrutinib but had owned the first molecule at Celera. It changed hands to Miller, who in all fairness rebooted the drug at Pharmacyclics although it didn't quite see success. And finally Hamdy, who had actively developed ibrutinib, had put his heart and soul into it and was on the verge of tasting success, was asked to leave before he could celebrate.

Ibrutinib had had a strong impact on people's careers it seemed. It was time for the molecule to make its mark on people's lives, the lives of the patients for whom the drug was created in the first place.

The wait was over.

"THE SECOND COMING"

December 2011
San Diego Convention Center

When Michael Wang took the stage to present the findings of a single-agent ibrutinib trial among MCL patients whose disease had returned, the massive ballroom was packed. Not a single chair was unoccupied; there was standing room only. The organizers had to stop folks from entering the room for it was quickly getting beyond capacity. I managed to get myself a place on the right side of the podium, close to the giant screen.

"Mantle cell lymphoma is an incurable cancer and among the more difficult lymphomas to treat," Wang began his talk. "There is no standard of care, especially for older patients. The younger patients are treated with very intense chemotherapy regimens, which come with a lot of toxicity."

Wang laid out the background and the need for better treatment options that were more feasible. He went on to present the data from the MCL trial: that of the patients enrolled across the globe by Wang, Rule, and Advani, among others.

The study had treated about fifty patients in total at the time of this ASH annual meeting in 2011, and almost all of them had very advanced stage 4 disease. Most of the patients were above sixty, something you expect in MCL. They had been previously treated with a variety of drugs, including chemotherapy, bortezomib-based treatments, and stem-cell transplants. At this late stage of the disease, MCL was even harder to treat, and responses to a new agent were rare. That's why the researchers waited patiently to learn about the response rates in MCL with only ibrutinib.

Wang was pumped up.

His voice cracked a little with sheer excitement as he was walking the audience through the patients' response data for ibrutinib.

"We saw an overall response rate of 67 percent, about 20 percent of them complete responders! This is unprecedented—better than any other single agent ever tested in MCL," he said. "We are witnessing a breakthrough—this is great news for the patients. The responses are lasting, with a long, progression-free survival."

To me in the audience, it sure felt like Wang was a savior. Here was this charming young doctor with a smile on his face presenting a magical new pill, taken by mouth, with no need for all the IV tubes, ports, and toxicities of chemo. And this doctor was hinting at a cure for a cancer that didn't even have a standard treatment option once the disease had progressed. Wang's talk was at the same time exciting and engaging, and paradoxically, it sounded utopian. I kept reminding myself that this was new data in a single-arm trial, but I couldn't help the excitement bubbling inside of me.

"They gave me a standing ovation," Wang recalled, years later. "I had to request the audience to let me finish my talk!" He always had an edge to his demeanor, and when it came to scientific presentations, Wang could be vivacious.

Simon Rule was seated in the front row at the San Diego Convention Center Ballroom, watching and listening to his colleague's speech.

"He's the born-again preacher. This was the second coming! We have a magic bullet, a pill that can cure cancer!" Rule thought.

Rule saw the talk as sensational, perhaps a bit outrageous—an observation I shared with him.

Yet the enthusiasm was not undue. Not many targeted treatments can even come close to what ibrutinib achieved in MCL in its first trial. This drug was transformational for this rare blood cancer. The patients who received it were simply overjoyed that a pill could reign in their out-of-control cancer, wouldn't interfere with their daily life so much, and allow them to carry on doing their job even. This was a dream cancer treatment for everyone involved: doctors, patients, nurses, and caregivers alike.

Ibrutinib was a direct challenge to chemotherapy. The question was: What were we going to do about it? Obviously getting approval was the next step, but then how do you make sure the patients have access to the drug? It's one thing to get a good drug for free through clinical trials, but once a drug's approved by the FDA, the patients need to pay for it. The cost of the drug can become a difficult hurdle to deal with.

"We were looking at a chemo-free world," Rule said. "I sensed right there and then, we would have a huge interest in and demand for this drug."

There were also psychological, interesting effects of ibrutinib among the patients. Even before doctors could document that the drug was working well and that patients were responding to the treatment, there were signs of improvement—at least that's what the patients conveyed. They took the pill and apparently immediately felt better.

I asked some patients who took ibrutinib and had a good response to it exactly when they started to feel better. "Almost the next day," was the answer from many. "I felt better immediately, on the second day, but did not want to say it then," one of the patients said. "You know, I didn't want to jinx it. It was my last chance." I knew Wang, Rule, Sharman, and others had similar stories to share.

The side effects were manageable with ibrutinib. Some patients had bleeding, dramatic bruising underneath their skin, but that really didn't bother them much. When you have overgrown lymph nodes packed with cancer cells in your neck area choking you, all you care about is to be able to breathe freely. All else comes second. Ibrutinib gave immediate relief by shrinking the overgrown lymph nodes, and that's what mattered most for these patients: a sufficiently controlled disease. Our friend David from Boca Raton and hundreds of others who has benefitted from this small molecule would share their feelings in the months and years to follow.

There were other side effects, though. Some patients developed a heart condition called atrial fibrillation (AF) while on ibrutinib. But only a few patients had it. Without the BTK drug, they would have received a chemo regimen. And with chemo, the number of patients developing AF is higher but people don't pay much attention to that side-effect profile of the chemo regimen. The same goes with bendamustine: the side-effect profile wasn't great for it either.

The *New England Journal of Medicine* published the updated, full data of the MCL patients in 2013. Wang was the first author and Rule the second. The paper reported a 68 percent overall response rate in a total of 75 patients whose disease had returned after an initial remission. The patient's response lasted on average for nearly eighteen months, and for more than a year there was no progression of MCL for these patients. These wonderful results cemented BTK as a golden target in MCL and set ibrutinib on course to be an all-time blockbuster drug in lymphomas.

Rule treated thirteen patients in the UK in the global trial. Save for patient number 13, everyone had a strong response to ibrutinib. He still feels bad for patient number 13. "The patient took himself off the drug. Everyone around him in the ward was high-fiving—all were responding to ibrutinib brilliantly, except him. It weighed on him. If he had continued taking the medicine, he perhaps would have benefited."

So the first win was achieved for team BTK: good response rates in relapsed patients. The next battle was already on the horizon: Could ibrutinib be given possibly as a first-line drug, right after the patients were diagnosed, in the older population who couldn't tolerate intensive chemo? Could it be as transformational there and perhaps cure the disease?

The answer might lie in the right combination of drugs and in a bit of serendipity.

It was a foregone conclusion: we knew ibrutinib would be approved in relapsed mantle cell lymphoma. The only question was if the drug would be available widely to meet the high demand.

On November 13, 2013, ibrutinib oral capsule received FDA approval through the accelerated pathway for patients with MCL who had received at least one prior treatment. Four months later it was approved for CLL patients with at least one prior treatment.

This heralded a new era in targeted cancer medicine, that a small molecule could be developed with so much speed in two different cancers and be approved pretty much simultaneously was pure joy for the cancer research community.

But the journey has just begun for the BTK inhibitors. We needed to bring this targeted treatment to our patients earlier in their disease, and ibrutinib presents a fantastic opportunity—for clinicians and drugmakers alike.

As a next step in ibrutinib's journey in MCL, the drug was planned to be given together with another non-chemo drug called rituximab with the hope that this combination would work better than ibrutinib alone. The study was known as a window trial at our institution (for once the trial naming made perfect sense to me) as there's a therapeutic window after initial diagnosis, when a new drug combination is given to the patients with the hope to induce remission of MCL.

GLOBE TROTTER

Summer 2012
Houston, Texas

The first thing you notice in the bayou city of Houston, or H-town as they call it, is that the land is as flat as it gets. A colleague of mine on the East Coast would joke, "The tallest summit you see there is the highway ramp. And you may need to drive for ten minutes just to get a cup of coffee!"

That proved to be exactly right. The enormity of the state of Texas can be daunting when you consider that ten hours of driving from Houston will only take you to the other end of the state, and even then Big Bend National Park stands in the way before you get into Mexico.

For a few months I was based at the University of Texas Health Science Center campus, literally next to the MD Anderson Cancer Center, applying for jobs like crazy and wishing for a position at a prestigious cancer institute. It was my dream to work at the MD Anderson Cancer Center, and my wishes came true on a fine October morning.

I received the all-important call from the Department of Lymphoma and Myeloma at the MD Anderson Cancer Center while running a PCR cycle in the basement of a university laboratory for a clinical research project I was working on.

"Dr. Ahmed, please report to the sixth floor of the faculty center for an interview with Michael Wang."

"Will do!"

The following morning I met Wang for the first time. If anything, he seemed to be a bit disturbed by the frequency of letters in my first name.

"What do I call you?"

By then I had been called many weird things so I was fully prepared.

"Call me Mak."

"All right, Mak, let's get to work."

Wang handed me a piece of paper on which he'd scribbled "window study and actionable genetic lesions in mantle cell lymphoma." I wasn't sure what this meant. Sensing my confusion, Wang said, "Overcoming ibrutinib resistance—that's going to be our primary focus."

The expected response of cancer cells to small molecule drugs such as ibrutinib was mutation—that is, a change in the coding inside the DNA of the cancer cells so that newer generations of cells were no longer inhibited by the drug. Apparently ibrutinib resistance had been reported in CLL. A threat to the effectiveness of the BTK inhibitor was emerging.

WINDOW OF OPPORTUNITY

At the MCL Center for Excellence at MD Anderson, we were a bunch of happy doctors, nurses, and scientists who soaked up the praise and pressure of the unprecedented success of ibrutinib in MCL. It was all about going into an earlier treatment line now—even the newly diagnosed patients—to give them the benefit of this wonderful drug. With Wang as our lead investigator, we designed this window trial.

It is a difficult decision to give a non-chemo drug to patients just diagnosed with an incurable cancer. The convention was to treat with chemotherapy since this had already been proven to work. But to push the boundary of therapeutics, we need to make these difficult choices. On one hand, the patients might respond to the new chemo-free drug and enjoy better quality of life with long remission of their MCL. But the flip side was that their cancer might progress more quickly. This posed a question: Was intensive chemo the better option? Initially the ibrutinib-rituximab (IR) combo was designed for elderly patients who were unfit for chemo, but later, the "window" concept included younger MCL patients.

Nonetheless, this window trial (which was called Window 1 since more versions of the window trials were coming up) took the bold step of giving newly diagnosed patients chemo-free drugs up-front—IR—as part 1 of their treatment plan. This was followed by part 2 treatment, which was a shorter round of chemotherapy regimen (hyper-CVAD). While the norm would be to treat these patients with eight cycles of chemo, we aimed to give only IR at the beginning, which was the window of opportunity to induce a remission, before giving chemo for only four cycles. The goal was to improve progression-free living for patients with MCL while reducing toxicity from the chemo regimen.

For a smaller cancer such as MCL (small because it's a rare cancer), this window trial created a lot of buzz. By then, Wang had already earned international recognition for the ibrutinib trial. Patients wanted to see him and receive an experimental combination in the hope of a cure. In only about a year we enrolled fifty patients in Houston for the window trial. The patients came from all over the country and even internationally (at least one patient traveled from China).

Our team treated six from Louisiana, and Bobby was among them.

RUN HAPPY

To know Bobby fully and to get into his perceptions of life and its trivia, you've got to know the cultural milieu that is Louisiana. Colonized by the French in the seventeenth century and settled by African, Spanish, and Caribbean folks until it was sold to America in 1803—Louisiana is a fertile ground for creativity, traditions, and folklore. True to his roots, Bobby was unpretentious with a deep southern voice, someone who took things as they came.

Yet he wasn't prepared for his diagnosis of MCL.

Bobby likes to call himself a golfer by trade. He plays with passion, something he has been doing since high school. Running comes second to him, but he only runs to keep in shape for his true hobby: golfing like a champ. But that doesn't mean Bobby would pass up a chance for outdoor fun. He had started prepping for a high-altitude run in Colorado with his son at one point, and that's when things started to feel different for Bobby.

Usually he could run six miles at a stretch, barely breaking a sweat, but now, even at five miles, Bobby started getting tired. He kept training to recover this higher level of activity, to stay in shape.

Unknown to him, he was getting in shape for a race for his life. Bobby shared his undue tiredness with his doctor, who thought they should evaluate the cause of this easy fatigue once Bobby was back from Colorado after the run. That's when Bobby's wife noticed a swelling in his neck.

"It was an enlarged lymph node on my neck, also another one on the back of my neck," Bobby recalled. He took antibiotics for two weeks, which didn't seem to work, and proceeded to undergo ultrasounds, a colonoscopy, and a biopsy of his lymph node. Everything so far looked okay—nothing suggested cancer up to that point.

Then the biopsy results were back.

The doctor came in with the biopsy results at hand. "He said I had mantle cell lymphoma." It wasn't clear to Bobby at first that this was a serious disease.

"MCL used to be very bad," the doctor told him. "But now there's a better survival rate." There were newer medicines that were out there. Some options were available locally. But years ago, Bobby and his wife had decided to seek medical help from nearby Houston at MD Anderson if either of them happened to suffer from cancer. This stemmed from the reputation MD Anderson had not just regionally but nationwide for treating cancer patients.

Bobby felt very emotional for a few days. "I have heard of other people who've been diagnosed. Never would I have dreamed that it would be me one day."

He wanted to feel better, trying to understand more about how bad the cancer was, but the sheer burden of emotion, the very thought that he had a growing cancer inside him, was just devastating to Bobby. He didn't know the staging yet, which was another source of his anxiety.

Soon he would. A barrage of additional tests and a bone marrow biopsy revealed stage 4 cancer. It was apparent that the MCL had spread to both above and below his diaphragm. Bobby's MCL also involved his bone marrow—another indication that the cancer was very advanced.

Because MCL is generally considered a slowly progressing disease, in Bobby's situation doctors tend to go for a watch-and-wait approach, waiting to see if the cancer progresses rapidly with no active treatment but close monitoring. But does it mean you wait for the dreaded symptoms? Wait for the inevitable progression when it can't be stopped anymore?

Bobby asked the question that thousands of other cancer patients have asked when they are diagnosed: "How many years have I got left?" This timeline of how long a patient has depends on whether the patient is in remission or not. In other words, once a patient achieves a remission of their disease, a timeline of survival (known as *progression-free survival*, or PFS) can be estimated. Without remission and just watchful waiting, the clock doesn't start ticking, or so it appears.

At least that's how Bobby took it. "I thought, if I could hold the treatment off, the timeline wouldn't start. Even though the cancer was there and in stage 4, the doctor seemed to think it could be years before they would need to treat me."

But things changed very quickly.

Bobby developed an ulcer in his stomach, and the local oncologist thought that it might have been caused by the lymphoma. About 80 percent of MCL cases involve the gastrointestinal (GI) tract, either upper or lower and sometimes both. In Bobby's case, signs of lower GI involvement were found as well.

The watch-and-wait approach was over. He needed to get treatment soon. At this point, Bobby decided to meet Michael Wang and our team at the MD Anderson Cancer Center in Houston.

Why did he choose a clinical trial over the standard treatment? What are the thoughts of the patients who take experimental drugs or new combinations?

To Bobby, it meant much that Wang led the trial and had written the treatment protocol. Wang had gained international recognition for his role in ibrutinib's approval. He saw MCL patients every day. In any other smaller center or city, the doctors seldom see MCL patients.

"They may open a book and say this is what the book says your treatment is!" said Bobby.

This perception was not far from reality. MCL is a rare cancer and not all hospitals or oncologists would be familiar with its new protocols and treatment choices. But for Wang and our team, it was a daily business to care for MCL patients.

Bobby was all set to begin the window trial. But there was something else he wanted to do before receiving the treatment. He wanted to run a half-marathon.

The marathon was scheduled to start in Nashville, about twenty miles from where Bobby's daughter lived. He'd signed up to run in the event a year before, and true to his Southern courage, Bobby wouldn't back off from running just because he had cancer.

"I was running to show that you can be sick and still run. You can be sick and still be active." Bobby persuaded us to allow him to take part in the half-marathon with a small caveat—his daughter would run with him just to make sure he wasn't overdoing it.

"It was a great experience to do that. My point was "Hey, we're going to do this and we're going to show this cancer who's in charge here."

On April 30, 2016, exactly fifteen days before his first treatment, Bobby completed his run in the half-marathon.

Bobby was in the first cohort of patients to be treated in the window trial. His MCL was in the advanced disease category, which included most of the patients in the trial. In that cohort of patients, everyone responded to IR: it took them only about two months to reach initial remission, and about six months to reach complete remission—a truly remarkable feat for a chemo-free induction regimen.

Bobby received his last treatment on March 3, 2017. He was in complete remission and so began routine, yearly follow-ups. Seeing his last clear PET scan, Bobby planned to run a full marathon with his son and daughter. Wang gave him his full support: "Enjoy your life!"

I believed that the chemo-free ibrutinib combination would be the new standard for patients like Bobby. For one, ibrutinib did something that no other drug in cancer (more precisely, in blood cancers) was able to do: it produced miracles, literally. There are hundreds of patient stories where they have been told to go into hospice because current practices and medical wisdom suggested there were no treatments available for them that could give any meaningful benefit at all. And yet they came out with remission of their cancer from ibrutinib.

I've seen elderly patients transferred from regional hospitals in Texas to MD Anderson for experimental care. Their only other option? End-of-life care. Yet I saw them getting complete remission of their disease after treatment with ibrutinib. When an eighty-five-year-old comes to you with a hospice referral and they achieve complete remission of their lymphoma, then walk out of the center with a bright smile and go on to live for more than five years with a good quality of life, what more can you expect from a cancer drug? It is truly sensational for those of us in cancer care and perhaps as close to a magic bullet as we might possibly ever be. Paul Ehrlich would have been genuinely proud of what targeted treatment such as the BTK inhibitors achieved.

What matters most to the patients is that they can tolerate targeted treatment like ibrutinib much better than chemotherapy. Logistically, a tablet or capsule taken orally allows them to be in the comfort of their home versus having to come to the hospital multiple times. That's why combination trials with ibrutinib are so attractive. In the age of COVID-19 these orally taken targeted drugs makes even more sense given that cancer patients are immunocompromised, and one less visit to the hospital means less risk of getting sick.

My colleague Maria Badillo, a fifteen-year senior registered nurse manager at MD Anderson, saw it all—from chemo regimens to targeted treatments only, to the window trial in MCL. She cared for hundreds of patients who received ibrutinib and heard their heartfelt gratitude. To her, the side-effects profile alone gave patients a lot of advantages over a conventional chemo regimen. "How can you expect a forty-five-year-old patient to accept neuropathy of their feet as a side effect of a chemo regimen?" Badillo says passionately. "It's not like the neuropathy is short-lived. It is not. It is a long-term problem; it's not going to go away. Yes, they may achieve good control of their MCL with chemo, but then these long-term side effects will need ongoing additional care and will cause significant disability. They don't want to live with that."

The accumulation of toxicities with chemo over time, the ongoing visits for receiving care and the need to take time off from work, park their car, find the money for their copay, and come up with a reliable caregiver to drive them to and from the hospital routinely are all very significant barriers for our patients. It is not that with ibrutinib all of these would go away—they still need to navigate some of these hurdles, but less intensive monitoring is a favorable factor for our patients when they are considering new treatments.

When a drug that is taken at home, with less-frequent follow-up visits and fewer intense side effects, it is always helpful to one's quality of life. And that's why targeted treatment is often preferred by patients.

But the question was: Could this new targeted drug give long-term remission? Could it possibly cure the disease, or were we taking a chance since chemo was a trusted treatment that typically gave long-term remission?

The complete data from the first MCL window trial later reported by our group showed the overall response rate was 98 percent for the chemo-free (ibrutinib-rituximab) portion only. More than 95 percent of the patients were living after thirty-six months of their treatment—a remarkable feat for our patients and their families. While we need more data, more follow-up to convince ourselves of the benefit of chemo-free regimen in aggressive newly diagnosed mantle cell lymphoma, it was clear that the novel targeted treatment is here to stay.

As for Bobby, last I checked, he was training his two grandchildren for a marathon.

THE COMPETITION: BATTLE OF THE "BTKS"

Currently there are six BTK inhibitors that are clinically approved and marketed for different cancers. We are now observing the battle of the BTK: there are second and third generations of BTK molecules out there, which followed in the footsteps of ibrutinib; which is going to conquer the throne (i.e., the market)? Which BTK inhibitor will have the best profile among all?

From a drug development standpoint, each of these BTK inhibitor is in a tight race to prove it is better than the others. The global market for BTK medicine is estimated to be about $10 billion today, and by 2030 this value could very well reach more than $20 billion. These are staggering numbers and, understandably, many drugmakers are aiming for a slice of this huge market.

This sets up another race for a remedy for drug developers working within the BTK space. If you are developing a class of drug that is already in the market and is being used by the medical community, you have much less room to do things differently than your competition. To a certain degree you need to follow a similar path of developing your molecule with a goal to set it apart from the rest, one way or another.

In this situation, we aim for differentiation—that is, addressing the question of how different the new BTK drug is compared to an earlier drug of the same class. The difference could be in its binding (covalent versus non-covalent), dose, how it's given (intravenous versus under the skin), how frequently it's given, or most importantly how much better the new drug is in safety and efficacy compared to others of the same class. This is a tricky task, proving one drug is better than another in a very competitive landscape.

There are multiple approaches to achieving this depending on the disease and the strategic direction of the drugmaker. One can choose to go for a head-to-head comparison—your drug versus mine, and let the best one win in a trial (a straightforward but costly and lengthy path); or, sometimes, merely proving that a newer drug is not inferior to the older one (older generation) might just be fine.

When it comes to differentiation, there are pricing concerns as well. The target for all these drugs is the same: BTK. Ibrutinib was a first-in-class BTK inhibitor, and so when it was first approved for use in MCL in 2013, the company could set its sticker price (before any discounts or rebates) at a whopping $130,000 per year! When other BTK–targeting drugs came to market (acalabrutinib, zanubrutinib, and others), their owners had to pay attention to their pricing to be competitive—especially since ibrutinib has already enjoyed strong sales.

Then begins the tug of war to see which drug can gain better market share, with companies employing various strategies to prove their drug is a better BTK blocker. Patients and prescribers now had the option to choose from several BTK inhibitors, so each company had its work cut out to demonstrate it had the best-in-class BTK drug.

In chronic lymphocytic leukemia (CLL), the BTK inhibitor drug class has cemented its position as a preferred choice (in combination with other chemo-free drugs) for newly diagnosed patients. For mantle cell lymphoma (MCL), I have been anticipating BTK inhibitors would become a standard of care earlier in the treatment as well. But this required strong data and a change of paradigm. The big question for me was whether we could expect to see chemo-free treatment given up-front in MCL when patients were newly diagnosed. Would this be considered too bold?

GOING FRONTLINE

June 2022
Chicago, Illinois

For the 2022 iteration of the American Society of Clinical Oncology (ASCO) Annual Meeting, I was more excited than ever because data would be presented that could help answer the question of whether ibrutinib, in combination with other drugs, given to newly diagnosed MCL patients could prevent or delay relapse of their disease. The study that addressed this question was dubbed SHINE (I still find it confusing how the naming of different trials work and why this name was given, not that it's a bad name).

The SHINE study had treated newly diagnosed older MCL patients with a combo of ibrutinib and two other drugs. Ongoing since 2013 after the approval of ibrutinib as a single drug in patients who had relapsed after initial treatments with chemo, SHINE had mature data: a long follow-up period to give confidence in its data on ibrutinib's long-term effectiveness. I'd also seen the early days of the SHINE trial as MD Anderson Cancer Center had been a participating site. This was a long-running trial that had apparently reached its maturity.

The presenter, on behalf of the SHINE team, was none other than Michael Wang.

After landing in Chicago for the meeting, I texted Wang, "Looking forward to your talk." I wanted to catch up with him after the presentation, to set up a meeting to hear more about the trial. As usual, he drew a big crowd. Wang was charming with his usual wit. But he was a lot calmer now, no longer the too-eager messiah and probably more careful than before. I guess years of experience and being in the spotlight had seasoned him well. He now delivered the data in a more grounded fashion, with his signature infectious smile.

"This is the first randomized controlled trial in newly diagnosed MCL patients who are at least sixty-five years or older," Wang began. "The trial examined the addition of ibrutinib to the standard treatment available already."

Most of our patients with MCL are men over sixty-five years of age. Unfortunately, because of their age and other health conditions, these patients are typically left out of clinical trials. Intensive chemo, or even newer targeted treatments, often have excessive side effects in an older age group, making them unfit for these treatments. For these patients, SHINE was a great option.

The trial treated half of the patients with ibrutinib in addition to the standard treatment, known as BR (bendamustine-rituximab), while the other half were treated with BR only (plus a placebo). For about six and a half years, the patients who received ibrutinib remained progression-free from their MCL cancer. On the other hand, those who received BR only remained progression free for about four and a half years. That's a gain of more than 2.5 years of progression-free survival for people who received ibrutinib.

Impressive results!

There was one caveat though: the long-term survival rate was the same for both groups. About half of the patients in the trial still passed away within seven years. So the study detected no difference in overall survival between the options. And survival is seen as the most valuable indicator as to whether a new treatment is better. Use of ibrutinib with newly diagnosed older patients did not outdo the standard treatment when it came to survival.

When a member of the audience asked if the results still mattered since there was no difference in overall survival, Wang was ready. "The study was designed to examine progression-free survival not overall survival. That would have been a much larger study." Unsatisfied, another audience member asked a similar question. Wang did not want to concede. He emphasized that the study participants were old. Half were over eighty. "So they are going to die from other causes!" Wang said. Although not incorrect, but this was a somewhat unsatisfactory answer.

The implication was that doctors would be taking a chance not giving up-front, aggressive chemo to younger, fit patients. But Wang had no doubt in his mind that ibrutinib should be the new standard treatment among older MCL patients.

"In my opinion, for newly diagnosed MCL patients who are older than sixty-five years, SHINE treatment protocol should be standard now," Wang concluded.

Not everyone was sold on adding ibrutinib to the standard regimen. With existing drugs such as BR, which were readily available and cheaper, the long-term survival was similar to what ibrutinib would achieve. And so it was tricky to uniformly accept ibrutinib–BR as the standard treatment, even for older patients. I have been involved in debates on whether the addition of ibrutinib in this older population is meaningful.

About one hundred cancer doctors who practiced in the United States met in the summer of 2022 in two live sessions to discuss ASCO presentations for the year. Of course, SHINE was a topic for discussion. Interestingly, more than two-thirds said they would still prefer to use BR alone due to availability and cost for newly diagnosed, older patients with MCL. However, the majority would consider adding ibrutinib to BR as a first treatment if the FDA approved this regimen.

As always, health authority approval is a key factor in choosing a new treatment. Ibrutinib had demonstrated benefit in patients who received other treatments first—there was no question about that. But as an earlier treatment option, it faces a higher level of scrutiny.

Johnson & Johnson (J&J), the marketer of ibrutinib in Europe, initially sought approval for using it in the treatment of patients with untreated MCL in the EU. But the European Medicines Agency signaled upon initial review that they considered the data from the SHINE study insufficient for such an approval. J&J formally withdrew its application in December 2022.

As for the United States, Pharmacyclics (now an AbbVie company) chose to voluntarily withdraw ibrutinib from the market for MCL patients who received at least one prior treatment. Again, the SHINE findings were not showing a clear benefit of adding ibrutinib early as a treatment option—so the prior FDA accelerated approval for the drug has not been confirmed. It doesn't seem like we would see ibrutinib move ahead in early (first or second line) treatment for MCL. Especially, as the patent coverage for the drug is expiring starting in 2027. But it is almost certain that other BTK inhibitors will seek to be the first course of treatment in MCL as a combination regimen even for younger, fit patients. It's a high bar to overcome, and it will be a watershed moment when it happens.

And so the stage for the next battle of the BTKs has been set. Six—possibly more—drugs are in the arena. The survival rates, efficacy, and side effects of each of these BTK inhibitors, and the cost to get the drug, will play a big role in deciding the winner, and the battle will be played out in the next few years, but patients will be the real beneficiary in this search for better molecules and better drugs. I am hoping we can switch to time-limited treatment with BTK drug class soon (a fixed-duration of treatment—not indefinite therapy until the cancer progresses) and identify an optimal medicine with higher efficacy and better side effects for the patients. There are ongoing and planned trials offering chemo-free treatments up-front, and I am looking forward to seeing the data presented soon.

Somewhat parallel to the ibrutinib saga, two other classes of drugs have made wonderful progress in cancer care since the human genome was decoded: monoclonal antibodies (a.k.a. biologics), and cell therapy. Both these drug classes represent the fast-growing, cutting-edge immunotherapy that has become a revolution in new cancer treatments, and I am fortunate to have been involved in some of these new drugs in development. As the BTK inhibitors race each other, let's dive into the saga of CD38–targeting antibodies.

PART II

A Story of Large Molecules: CD38 Antibodies

Chapter 6

The Unmet Need

Regardless of our many differences, we all have the same needs. What dif-
fers is the strategy for fulfilling these needs.

—Marshall B. Rosenberg

TO OLDE TOWNE

July 2018
Houston, Texas

It was at the end of a grueling funding-review season for the Cancer
Moonshots Program at the MD Anderson Cancer Center. Recuperating from
a fractured left tibia and my own cancer scare, I wasn't feeling good but
nonetheless decided to show up at work on crutches.

Michael Wang saw me at the end of the hallway, at the entrance to our
department at the faculty center. He looked cheerful, as usual.

"I must say, you are holding up better than I expected," Wang said. He was
aware of the situation with my biopsy—we had spoken about it before.

I hopped on a golf cart that transported people between the buildings con-
nected by covered passageways in the massive MD Anderson campus. Wang
joined me for the short ride. We were headed to a project review session for
Cancer Moonshots in the West Building.

The MD Anderson Cancer Moonshots Program gets frequently confused
with the moonshot announced by President Obama in 2016. Since I joined the
program there had been numerous occasions when I had to explain to patients
and caregivers that the two moonshots were not the same!

In contrast to the federal initiative, the MD Anderson Moonshots Program is
an institutional effort launched in the fall of 2012 to accelerate new treatment

discovery for several deadly cancers and to improve our patients' survival rapidly. The program allows the participation of patients who have cancer and brings in pledges for them however big or small the contribution may be. By 2016 when the federal moonshot was announced, the MD Anderson program already had commitments of more than $300 million from altruistic donors who were keen on helping create breakthroughs in cancer care.

I was heavily involved in the small molecule clinical projects, including chemo-free trials in MCL. Our efforts in MCL had now branched into overcoming drug resistance in lymphomas and understanding how MCL cells escape therapeutic pressure through mutations. Another category of drugs known as large molecules (also called *biologics*) was big in several cancers, and featured in the moonshots project proposals. Biologics are complex forms of medicine usually reserved for the treatment of cancers and immune-related diseases.

A key difference between a small molecule such as ibrutinib and a biologic drug is that small molecules are created chemically while the biologics are obtained from living cells. Biologics are costly to manufacture and unlike small molecules cannot be given orally. These drugs are either injected or infused over a longer period. Large molecules are one of the most versatile drug classes that currently exist.

Several projects in the moonshots program were in collaboration with pharmaceutical companies because we needed new drugs from them to continue doing innovative clinical trials and testing these molecules in new research projects. I felt good about the data we were generating in collaboration with the world's top drugmakers.

But by then I was considering a move to the pharmaceutical industry having seen the practice-changing scope of work there. In academia, one needs to be a top-level expert to influence practice and of course secure grant money. But the reality of it is that most clinicians, scientists, and practitioners are not key opinion leaders. We are merely followers. There is no problem with that until one realizes that the opportunity for drug development is rather narrow in this capacity.

Primarily, we enroll patients in a clinical trial per the protocol written by our colleagues on the industry side. The development of a molecule, its directions, monitoring, and major decisions involving it, are largely controlled by the drugmaker who owns it. If someone aspires to design and develop new medicines that could change the therapeutic landscape, the drug-making industry is where they would be more impactful. This industry is at the forefront of the new era of targeted medicine we are in. Pharma has the finances and business capacity to make things happen: to develop new drugs with the participation of patients, caregivers, doctors, hospitals, and the government, and so I was inclined to explore a new direction for my career as a drug developer.

Later in that scorching July afternoon, I had to stop at the orthopedics office for an appointment because my fractured tibia was hurting badly. I had one more meeting on my calendar, and one I was nervous to take. It was a conversation with a pharmaceutical industry expert who had invited me for an exploratory chat.

Though I wasn't 100 percent sure what they wanted to speak with me about, I had a sense it could be about a position in the industry for drug development.

I spoke with them from my parked car, and it ended up being a very consequential thirty minutes for my career. They wanted me to lead the development of new molecules in blood cancers. It would be different from my current work because the job would be on a global scale.

At first it sounded exciting and, honestly, better than filing grant applications tirelessly and worrying about the next cycle of applications. If I accepted the role, I would be able to take part in the drug-development process actively and more closely than I was ever going to be able to do as a physician-scientist.

As I thought about it more, the practicalities of changing the career trajectory started to weigh on me. As a physician and a cancer researcher, I had been in academia all my professional life so joining a pharma company would be a whole new environment. But deep inside I knew this was where I could use my skills to directly influence the creation of anti-cancer agents. It felt worth the many personal disruptions in our lives (three young kids, including a newborn) relocating across the country to the Northeast.

In the winter of 2018, we moved to Boston—the Olde Towne. There, unlike my work in small-molecule-targeted therapies, I would start with the development of a large molecule that relied on a much older healing modality discovered not long after Massachusetts ratified the U.S. Constitution.

Time would tell if I could harness it to make my own breakthrough.

Chapter 7

A Very Unusual Matter

Then we should find some artificial inoculation against love, as with smallpox.

—Leo Tolstoy

THE BIRTH OF A VACCINE

May 1796
Gloucestershire, England

Edward Jenner loved poetic expressions.

Despite being a popular doctor and a scientist, he couldn't help explaining his observations of diseases and their treatment in verses whenever he could. To Jenner, science became more relatable once the narrator actively tried to simplify his or her explanations. His father was a vicar in Berkeley whose wife, Sarah, had nine children, of whom Jenner was number eight. Perhaps his life-defining learning experiences came from the medical training he received under the supervision of the famous Scottish surgeon John Hunter. Hunter was a pioneer experimentalist, anatomist, and natural scientist.

In 1770, Jenner came to train with Hunter at St. George's Hospital in London. By the end of the year, Hunter already knew that his young protege had extraordinary abilities in dissection and experimental investigation. The two men became not just mentor and mentee but lifelong friends and collaborators. But there were striking differences between the career and life paths chosen by the doctor duo.

Gifted with superior mental qualities and knowledge, and despite his deepest respect for Hunter's work, Jenner chose the somewhat quiet life of a rural doctor. His chosen path did not have the prestige of a medical instructor in

London, like Hunter, nor did it have the excitement of a naturalist. In 1772, none other than Captain Cook invited Jenner to take part in his famous expedition to Australia. For an adventurer, this was the voyage of a lifetime: to study natural sciences and to uncover the unknown world. But Jenner rejected the invitation, choosing to remain a rural practitioner.

Today, history recognizes Jenner as the father of vaccines—an essential tool in triggering an effective immune response in the body against threats by making the practical application of antibodies possible. This was one step closer to a future of using the body's immune response to fight cancer, which is called *immunotherapy*.

Completing his training in London, Jenner began his practice in Berkeley in 1796. He spent months thinking about the idea of an acquired immune response to smallpox. As a child he had contracted the virus and survived. Jenner was much intrigued by the folklore that dairymaids who had caught cowpox from milking infected cows escaped smallpox with only mild symptoms.

Back in those days there were no medical journals. Local doctors would discuss their own observations in social gatherings. Young Jenner may have picked up this folklore from the time when he was an apprentice in a society of clinicians in the region. He knew in his heart there had to be a simple explanation for this immunity. He was obsessed with cowpox infection and its supposed protection against deadly smallpox.

In the spring of 1796, a young dairy maid, Sarah Nelmes, came to see Jenner with three lesions on her hand. She was worried that they could be smallpox. Upon examination, Jenner determined the lesions were from cowpox that she had contracted from a cow of her master.

Jenner's mind raced. Nelmes's lesions presented a great opportunity for him. He wanted to generate experimental proof that prior cowpox infection prevented patients from getting very sick with smallpox. What he planned to do was very similar to something called *variolation*. Variolation would mean exposing someone who had never had smallpox before to smallpox by taking smallpox material from an infected person's lesions and then exposing the smallpox-naive person to that material. This was a common preventive technique among the doctors in the 1790s.

But who would be Jenner's subject in this experimental use of cowpox?

He summoned James Phipps—his gardener's son and eight years old at the time. Phipps became the first subject of a vaccination experiment in recorded history.

Jenner took cowpox infected matter from the dairy maid's sores by cutting open the pustules with the sharp point of his scalpel. "The matter was then inserted into the arm of the boy by means of two superficial incisions, barely penetrating the skin, each about half an inch long." Jenner described the inoculation process neatly in his 1798 *Inquiry*.

One can see the ethical problem with such an experiment. And frankly, when you consider that Phipps was the child of an employee of Jenner's and there was a major risk of his getting sick with such a method, this experiment probably wouldn't have been approved by a research ethics committee if any had existed back then.

Then again, because the experiment was very similar to variolation, a standard practice, it had some justification.

Questionable as the procedure might have been, Jenner succeeded in awakening the boy's immune system.

"On the seventh day, the lad complained of uneasiness in the axilla, and on the ninth, he became a little chilly, lost his appetite and had a mild headache," Jenner wrote.

But the boy's symptoms had improved by the following day, and the incisions did not show any signs of progression. Six weeks later, Jenner exposed him to smallpox material through variolation, but Phipps was now resistant to the smallpox. In the succeeding months Jenner exposed the boy at least twenty times to smallpox but he was never infected by the virus.

Jenner's vaccination experiment, the first ever in the history of medicine, was a complete success. He had succeeded in vaccinating James Phipps against smallpox through Phipps's earlier exposure to cowpox. And yet, despite his intentions, Jenner did not publish his results at this stage. Two years later in 1798 he eventually published his vaccination results in *Inquiry* after repeating the experiment with additional subjects.

The cowpox infection gave perfect immunity against smallpox, and to Jenner it offered a much safer alternative to variolation. Jenner couldn't provide the reason for the success of cowpox exposure though. Maybe he found the process so befuddling it dulled his usual wit. Jenner was a self-proclaimed procrastinator. He alluded to the idea of a "matter" that was not identical to the virus but something that was generated inside the human body in response to the virus by "some peculiar process."

We now know that this matter is a protein secreted by B-type white cells in our blood in response to an invader that the protein has an affinity for. Thanks to our friend Paul Ehrlich, we now commonly refer to these proteins by his name for them: *antibodies*.

As for the "peculiar process," after an infection by the cowpox virus, the human body has the ability to detect the similar smallpox virus's antigen and is able to fight back because the two viruses are closely related. When an antibody attaches itself to and flags a cell, it rallies the immune system to attack and destroy invaders like it. The ability of antibodies to mount a powerful counterblow against an invader requires a very specific affinity between the antibodies and the invader generating those antibodies. That's why the invader is dubbed an *antigen*—short for antibody generator.

But since antibodies are produced by a living biological system, their behavior is unpredictable and not always sufficient to destroy the invading cells. This is especially true if the invader is a cancer cell. Cancer cells can mask their identity, avoid detection by the immune system, and even confuse immune cells by blocking their messaging network.

Nonetheless, cross-immunity was at work for Jenner, who predicted the ultimate success of his experiments. "The annihilation of the smallpox, the most dreadful scourge of the human species, must be the final result of this practice," he wrote.

While annihilation sounds unrealistic for our current cancer scourge, in the late 1700s it was equally unrealistic for smallpox. I loved the way Jenner confronted smallpox and wanted to have some of this boldness on my own path to fighting cancer.

Jenner's prophecy happily came true nearly two centuries later in 1977 when the World Health Organization reported, in Somalia, the last natural case of smallpox—an outstanding achievement for medicine and immunology.

But awareness of this "peculiar process" and its potential was not enough. The world needed to wait for another one hundred years before Paul Ehrlich would begin to dream of the magic that could be accomplished with this protein and feel inspired by the alchemization of rabid animals that would allow him to find a way that actual treatments could be created.

1890
Freiburg, Germany

Paul Ehrlich believed that scientists needed to learn to "aim chemically." He theorized that drugs (which were mostly chemical agents in the late 1800s) enter the human body and immediately find their target in the cell. So if a chemical could attack a pathogen precisely, the healthy tissue wouldn't suffer from the actions of it. Through his early works in dye staining, he knew all too well specificity was the key. If chemicals were tweaked, they could very well hit a disease-causing agent and heal without causing toxicities.

The magic-bullet (*Zauberkugeln*) concept was born. This cure by chemical doctrine still captures our imagination to this day. On more than a handful of occasions we've declared miracle cures for cancer over the past decades and in some cases bestowed the magic-bullet accolade on these medicines. But the reality of it is that we are unlikely to find one magic bullet for cancer even though the concept of targeted treatment still very much resonates with that hope.

Despite his brilliant and incessant work, Ehrlich had powerful opponents in the scientific arena. Some derided his experimentation process as ruthless or labeled his work as a waste of money and resources—much to Ehrlich's irritation. In this pioneer's mind, controversy and perceived doubts about his work were sterilizing and took away the valuable time needed to prove things he considered already settled.

Nonetheless, Ehrlich continued to focus on his own brand of research—a blend of chemistry devoted to furthering the knowledge of biology. He was no one's disciple. He taught himself and worked in the laboratory with his own hands, limiting the number of people in his lab. "Avoiding many doves flying in and out" was his mantra.

At that time, Emil von Behring was an assistant scientist to Robert Koch at the Institute for Infectious Diseases in Berlin. He and Japanese physician-scientist Shibasaburo Kitasato, who had also studied under the supervision of Koch, demonstrated that diphtheria and tetanus toxins could be neutralized by antitoxins in animals. Essentially, toxins are poisonous substances taken from plants or animals. Antitoxins are antibodies generated in the body as a response to the toxins because they can neutralize them.

Von Behring collected blood from immunized rabbits that had survived a tetanus infection. The blood was allowed to settle until *serum* (the clear top portion) was formed. This serum was then transferred to the abdomen of uninfected mice via injection. These mice were later exposed to tetanus toxins but the animals never developed tetanus. What's more wonderful was the discovery that the serum collected from the immunized rabbits could be administered to animals suffering from tetanus and it would rescue them in turn!

With Kitasato, von Behring conducted similar experiments for diphtheria toxins, and the scientist duo successfully created diphtheria antitoxin. The world came to understand that serum from immunized animals could both prevent and treat infection. Treating diphtheria and tetanus is a great example of the power of antibodies.

But a question remained: Was there a reaction between the toxin and antitoxin that neutralized the toxin, or were there biological factors at play inside the living animal when these experiments occurred? Other interactions inside the living body could also render the toxin null.

Ehrlich had a hypothesis.

He had collected proteins from poisonous plants like the castor-oil plant, which has beneficial oil but a poison called *ricin* in its seeds. If animals or humans ingest the seeds, fever, cough, a tightness in the chest, and serious breathing distress can occur. For his experiments, Ehrlich injected ricin toxins in animals. This resulted in an antitoxin—anti-ricin—in the animals' blood.

In the first-ever test-tube experiment in immunity studies, Ehrlich then collected anti-ricin containing serum from these animals and added red blood cells to it. Normally, ricin toxins would clump red blood cells and destroy them, but Ehrlich had anticipated this, and because of the anti-ricin from the animals' serum, the poisonous ricin toxin did not clump the red blood cells. They remained discreet, doughnut-shaped structures, which is the normal state of red blood cells when they are not under stress.

His predictions were correct.

Ehrlich concluded that antitoxin acts directly on the toxin and neutralizes it. Just like his dye-staining principles, he posited that toxins and antitoxins had a special chemical affinity for each other. Through this attraction, one canceled out the toxicity of the other.

Although von Behring and Kitasato showed that diphtheria and tetanus could be countered by antitoxins in a living animal, it was very problematic to obtain and prepare these antitoxins to be given to a human being as a treatment.

Robert Koch invited Ehrlich to his institute in 1890, and there Ehrlich started working alongside von Behring. He proceeded to champion a method to standardize production of antitoxins using horses for large-scale production of antitoxin serum. Von Behring and Ehrlich formalized their partnership in creating serum therapy by forming a laboratory under a railroad circle in Berlin.

Soon, a new institute was founded, and Ehrlich was appointed head of the Institute of Serum Research. Von Behring's role as a principal discoverer of the serum therapy won him the Nobel Prize for Medicine in 1901, but thanks to Ehrlich's biological standardization methods, these therapies could actually be used to treat diseases.

Seven years later, the Swedish academy awarded Ehrlich his own Nobel Prize in 1908.

Despite the standardization of serum production, a practical challenge for Ehrlich and other scientists at the time was how to generate enough antibodies consistently so that they could be used to treat diseases. In other words, how best to stimulate the immune system to create a specific antibody in large enough volumes that could be used as a treatment? Natural immune processes were too varied and low yielding. For therapeutic application of serum, it must contain an antibody that is specific for the disease-causing microbe; it must have sufficient antibody to fight the disease; and it must be given early in the course of disease when symptoms start to show in a patient. Aside from having to use immunized animals for production, naturally occurring antibodies lacked specificity and exhibited variability in binding to its antigen.

These problems limited production of single origin antibodies as a therapy for decades, until an Argentine biochemist and his colleague mastered a new technique.

But a lot more would need to be known about these magic bullets first.

THE BOY FROM THE BAHIA BAY

1935
Argentina

About five hundred miles south of Buenos Aires, Bahia Blanca is a port town on the shores of the Atlantic Ocean. At the turn of the twentieth century, Jews of Ashkenazi descent started to migrate to Bahia Bay, wanting to settle in a small town for farming opportunities. Many were escaping persecution in Eastern Europe.

This is where our next protagonist in the antibody saga, Ciser Milstein, was born.

Milstein very much liked to play on the dusty streets of the settlement as a child. His parents were of Ukrainian heritage, but unlike those who came to the port town for farming and labor, his mother, Maxima, relished intellectual pursuits.

His mother's encouragement didn't seem to work at first as Milstein didn't quite like reading.

That changed with *The Jungle Book.*

Kipling's adventure stories completely fascinated Milstein, and especially the freedom and laws of the jungle. This reminded him of the streets he played on. The settlement and its streets were a fantastic world—here young Milstein had a sense of freedom, a deep connection to his surroundings. There were moments of pure joy in running around exploring his world and disappointment over having to get back home in the evening. Playing with other kids was the highlight, but he was keen to know all that was happening out there. The world had so many wonders to unpack for little Milstein—just like Mowgli's world in *The Jungle Book.*

But it wasn't until 1935 that Milstein began to envision how he might gain access to that fascinating jungle. A cousin of Milstein's came to visit the family after graduating with a degree in chemistry. He was working at the Malbran Institute in Buenos Aires as a biochemist.

When eight-year-old Milstein asked him what was most interesting at his work, the cousin said, "Extracting venom from snakes."

Young Milstein was very surprised.

"What?"

"Yes, we collect venoms from snakes, and use it for research," replied his cousin. He explained to Milstein that extracted venom was used for treating snake-bite victims. This resonated with Milstein. He knew the jungle was full of exotic animals that lived in harmony, supporting each other. He became interested in science. He wanted to learn more of this intriguing world of discovery.

But what really determined his fate was what Milstein got for his ninth birthday present from Maxima: *Microbe Hunters* by Paul de Kruif. Milstein cherished the adventurous lives of legends like Louis Pasteur as narrated in *Microbe Hunters*. He wanted to be like Pasteur.

So Ciser Milstein went to study at the University of Buenos Aires.

In the mid-twentieth century, scientists were beginning to understand the structure and shape of an antibody. Antibodies are shaped like a *Y*, having two protein subunits—one light and one heavy chain—held together by a *disulphide bond*. The stem of the *Y* was considered a *constant region* while the tips of the arms of *Y* were labeled *variable regions*. A puzzle had remained unsolved for decades: if antibodies all have a basic stable formation containing the same proteins, how could they target multiple groups of antigens at the same time? This question on the diversity and specific attraction between antigens and antibodies had intrigued the pioneers like Ehrlich in the 1900s.

Milstein, though, was not focused on antibodies at this point.

In 1957 he won a best doctoral research award in chemistry offered by the Association Quimica Argentina. It allowed him to publish his works in several scientific papers and gain visibility. While his thesis was unrelated to antibodies, focusing instead on investigating an enzyme called *dehydrogenase* that catalyzed removal of hydrogen atoms from other molecules, Milstein's research also addressed the strong bond that provided stability to protein structures. This bond was the disulphide bridge, the same bond seen in the antibody structure.

His career began to take on a new course when Milstein met Frederick Sanger, a new Nobel Laureate in Chemistry in 1958, who bonded with the man from Bahia Bay. Sponsored by the British Council, Milstein joined as a fellow at the University of Cambridge, Sanger's workplace, just two weeks after he had won the coveted prize. Although Milstein returned to Argentina after the fellowship, Sanger eventually convinced Milstein to join him at the Medical Research Council (MRC) Laboratory of Molecular Biology (LMB) in the UK to work on protein chemistry. Political turmoil in Argentina played a big part in Milstein's decision to move to the UK.

In Cambridge, he started to investigate a specific protein, Bence-Jones protein. The abundance of this protein in the urine and blood of patients is a feature of multiple myeloma (MM)—a cancer of plasma cells. Bence-Jones protein has the same structure as the light chain of an antibody so it offers a key tool for understanding the formation and diversity of antibodies. The question Milstein wanted to tackle was this: How does a stable antibody structure fend off such a diverse range of antigens?

Milstein thought the diversity of it might arrive from the disulphide bridge, the stable stem of the antibody's *Y* shape. This was the region where antibody chains were combined, and Milstein said it offered "my own short-cut to the understanding of antibody diversity." The two sulfur groups, joined by a linkage and denoted *S-S*, represented a common architectural feature of a variety of classes and subclasses of antibodies, yet the distinction between different antibodies could come from diversity in this *S-S* bond.

Diversity stems from variations or mutations in genes that code for proteins. Now, there are two types of mutation: *germline*, which occur in the sperm or egg, the kind of mutation we inherit from our parents; and *somatic*, which happen in cells other than the sperm or egg after we have been conceived in the womb. Somatic mutations cannot be passed on to future generations so they are not inheritable. These mutations can be caused by exposure to chemicals, radiation, or viruses or even by accidental errors during DNA replication. Today, much effort in cancer research is focused on understanding the role of somatic mutation in causing cancer, escaping therapies, and on targeting such mutations with drugs.

Milstein hypothesized antibody diversity might be generated by somatic mutation as opposed to germline mutation. But he had a lot of difficulty testing this idea—that is until, for once, cancer proved to be a hero.

THE MONOTONOUS PROTEIN

Ever since the start of the twentieth century, scientists had had trouble isolating and purifying single-origin antibodies. Although there are billions made each day in the human body, obtaining one reference antibody in sufficient quantity was hard. But in scientific experiments, one needs a standard (a control) to compare with other variants.

In our body, plasma cells are a type of white cell that produces antibodies. These cells develop from B-type white cells and their secreted antibodies are transported by blood to the site of antigens where they work to neutralize a threat. Usually, activation and antibody production in plasma cells is orchestrated, normal plasma cells produce a diverse collection of antibodies in every healthy person.

But in MM, large numbers of plasma cells start to grow out of control in the bone marrow where blood cells are produced. These plasma cells crowd out natural blood-forming cells and spill over into other parts of the body.

In 1951, American immunologist Henry Kunkel recognized that myeloma cells have the unique property of producing a single antibody in large quantities in contrast to the vast array of antibodies that healthy plasma cells produce. Kunkel and his colleague at the Rockefeller Center in New York began to collect myeloma protein (a.k.a. Bence-Jones protein) from the blood or urine of patients for investigating normal antibodies. This monotonous protein showed up as a spike in electrophoresis graphs and was called an *M-spike* (short for monoclonal spike).

For Milstein's research on antibodies, Bence-Jones proteins proved to be vital. The researchers needed a supply of reference antibodies to test, and this protein from MM patients gave them just that.

A real breakthrough came in 1962. American molecular biologist Michael Potter at the National Cancer Institute in Bethesda discovered that injection of mineral oil or plastic material in the abdomen of a specific strain of mice led to the growth of myeloma cells in the animal. So by introducing these substances to the mice, he could ensure a stable supply of myeloma cells in the animal—finally, a source of antibodies other than the blood or urine of myeloma patients. This was a lucky discovery for Potter, whose laboratory became a fount of myeloma cell lines. Researchers around the world desperately needed this for their experiments.

Despite the groundbreaking work of Potter, these myeloma cells still needed to be collected from living mice, which was complex and time-consuming. In 1970, scientists at the Salk Institute in San Diego made it possible to generate myeloma cells outside a living animal. They found a way to use Potter's myeloma cell line in a suitable tissue culture medium. In the culture, these cells could grow indefinitely and secrete the myeloma protein in large quantities.

The problem of getting a single origin antibody in sufficient amount from a tissue culture was now solved. Ironically, immunotherapy treatment for cancer was greatly indebted to a cancer (MM) itself. Without this myeloma protein, antibody research would have been stalled.

In Cambridge, Milstein obtained Potter's myeloma cell line. This was a huge boost for his research, and he went on to acknowledge this in his Nobel lecture later in 1984.

Milstein finally had a way to test his original idea that antibody diversity might be generated by somatic mutation. The aim was to detect somatic mutations in Potter's myeloma cell lines and piece together the relationship between the mutations and the diversity of the antibodies.

With this goal, Milstein and his colleagues painstakingly worked for three straight months. But from a whopping seven thousand clones of myeloma cells, only five variations of the antibody structure were identified.

The reality is that somatic mutations rarely happen. The team found out that mutations happened on their own in the myeloma cell line without any modification of the culture medium. What was more revealing was that the major mutant antibodies found in nature did not match what was identified in the cell lines in the lab. The natural antibodies appeared to be following a different adaptation than the ones in the laboratory, another proof that cell lines don't totally resemble the cells they've been created from.

Milstein was back to square one. His hypothesis of a mutation-prone segment for the antibodies couldn't be proven. That's when Georges Kohler decided to join the team.

FUSING ONE CELL INTO ANOTHER

Dick Cotton worked as a postdoctoral fellow in the Milstein lab in Cambridge, UK. He had an idea: What if we fuse two myeloma cells together?

Initially it sounded interesting but not important to Milstein. But the team needed to do things differently. Milstein okayed it and Cotton fused two myeloma cells (one from a mouse and the other from a rat) through a chemical process. Typically antibody producing cells use only one set of genetic instructions from one parental cell. This implies that in antibody production, plasma cells silence one set of parental genes while the other set is preserved. This contrasts with the majority of cells in our body, which inherit both sets of genes from parental cells.

In fusing the two cells, Milstein and Cotton wanted to observe which genes were transferred or silenced.

The results of the cell fusion surprised the two scientists. The fused, hybrid cell expressed genes of both parental cells, not preferring one over the other. While this was not what they were looking for, the results allowed Milstein and Cotton to offer a theory that the variable and the constant regions of antibodies might already have been determined before the plasma cells were mature enough to produce antibodies. This also offered a hint: Could fused myeloma cells retain the properties of both parental cells?

Milstein presented these findings of myeloma cell hybrids at a seminar in Basel. The data intrigued Georges Kohler, who also wanted to explain somatic mutations leading to antibody diversity. He was a PhD student at the time at the University of Freiburg. In 1974, Kohler joined Milstein's lab in Cambridge as a postgraduate scientist to work on this further.

Cotton's fused myeloma cells were a good start, but Milstein and Kohler wanted to fuse two cells that would produce antibodies continuously. Kohler and Milstein thought if they could fuse a normal B-cell (which produced a specific antibody) with a myeloma cell (which would be immortal as all myeloma cells are), the resultant fused cell would have two very desired properties: the abundance of a specific antibody (from a B-cell) and an indefinite lifespan (from the myeloma cell).

Kohler set up the experiments. By January of 1975 he had fusion cells ready for testing. The hybrid cell—called *hybridoma*—secreted a large volume of antibodies in the laboratory. Milstein and Kohler repeated the experiments twice more just to make sure, and the same results were found.

The team now had an immortal hybrid cell line producing a single antibody continuously.

"The lack of success led to our breakthrough," Milstein said. "Since we could not get a cell line off the shelf doing what we wanted, we were forced to construct it."

The method of fusing two cells came to be known as hybridoma technology. The antibody it produced was called a *monoclonal antibody* because a single lineage (clone) of cells secreted the product. This technique allowed creation of antibodies in the laboratory that could be made against a specific target antigen. It was in August 1975 that this groundbreaking discovery changed the field of antibodies and the medicine industry forever.

A new era had begun.

WHEN SCIENCE MEETS COMMERCE

A firestorm of controversy broke in the late 1970s regarding Milstein's monoclonal antibodies. It wasn't about this new technology's science or its use but something I would need to become acquainted with as a drug developer: its commercial viability.

Britain was observing a sharp decline in manufacturing, and unemployment rates were rising. A therapy from Milstein and Kohler's monoclonal antibodies would be a financial boon for the country, considering the high value of new treatments. As it turned out, the commercial potential of hybridoma was unrealized.

For research purposes, demands for Milstein and Kohler's antibody skyrocketed after 1975. Scientists across the world requested their hybridoma. By the end of 1977, Milstein himself had shared the hybridoma cell line with at least fifty researchers. Transporting cells became a big headache—in-person collection, test tubes in individuals' pockets, postal services, via trains and planes, it was a total mess. It's impossible to imagine sharing or transferring technology this way in today's world.

Very soon it became a futile operation for the research group at Cambridge. The team couldn't keep up with all these requests for hybridoma.

In 1977, David Murray, founder of Sera-Lab, offered to distribute hybridoma on behalf of Milstein on a fee-for-service basis, returning a slice of the profits to his lab. Milstein was relieved to have a way to outsource the mailing, sharing, and overall distribution of his invention.

Bizarrely, the Milstein and Kohler hybridoma method of antibody production was not protected by a patent with the Medical Research Council's authority, the National Research Development Corporation (NRDC) in

England. NRDC officials did not "identify any immediate applications" of the Milstein and Kohler method and so no patent was filed.

Margaret Thatcher, the Iron Lady of England, did not like the fact that Milstein's government-funded project had failed to file a patent. Thatcher was a chemist by training, and she saw a missed opportunity for securing a British economic advantage through Milstein's invention.

In today's clinical research, the system ensures that such mistakes do not happen. The commercial scope of medicinal products is huge, and as financial value increases, the protection of intellectual property becomes crucial. As a drug developer, I must constantly keep in mind a drugs patent's expiration period even before that drug is formally available! And this applies to researchers in academic settings as well. They too are very much aware of the need to protect intellectual property.

When a new prescription drug is created and has completed its required trials, the manufacturing company launches the product. At launch, the company has the exclusive right to develop, sell, and market their medicine for a limited time. This period varies, but for cancer medicines, it typically lasts ten to twelve years. Once this period is over, the company has to relinquish control of the drug's formula to the marketplace—something known as *loss of exclusivity* (LOE). Once your drug reaches LOE, others can produce and sell generic versions, that is, copies of the original, brand-name drug.

It is kind of like the *Game of Thrones* saga: kingdoms (pharma companies) compete for control of the iron throne (the market) while other noble families (biotechs and startups) are on the lookout to sell their assets to the big fishes before large investments are made. And of course, winter (LOE) is coming!

It's a race for remedies. What is relevant today and an attractive potential drug may not matter merely months afterward because someone else has already started developing a similar drug. A tsunami of molecules, small and big, are making their appearance every day, and a drug developer must be aware of what's happening in the pharma kingdom.

Absence of patents for Milstein and Kohler's hybridoma methods led to an opportunity for Hilary Koprowski, Carlo Croce, and Walter Gerhard, who filed and were granted patents for monoclonal antibodies in the late 1980s. The patents used Milstein's original cell line, and many scientists thought of this as profiting from someone else's work.

An investigation into the matter in 1980 put much of the blame squarely on the scientists behind the invention: Milstein and Kohler. The investigators reported that the scientists' "lack of awareness" led to the fiasco. This irritated Milstein.

An important issue in the 1970s was that scientists were not entitled to royalties from patenting an invention in England if their project was funded by the government. If the policy allowed royalties for the inventors, would

Milstein have paid more attention? I don't think he would have acted any differently, but nonetheless, this could have been the case.

Looking back on the patent saga later in life, Milstein considered it a blessing that his methods were unpatented. It allowed him the freedom to keep doing antibody research and publish in scientific journals, which would have been restricted had the NRDC filed a patent. It also allowed Milstein to share his antibodies with other scientists and help advance science.

While I agree in the spirit of science to share findings, to allow innovation to make a significant impact, I also think inventions need to be protected. This ensures future investment in science and rapid advancement in therapeutic research and support growth.

In 1980, MRC finally changed their policy to allow a portion of any resulting royalties to go to the inventors. Some would say it always should have been this way.

As for Milstein, he did not care about the missed proceeds he might have received had the patent been filed. His response to this was, "I was not unhappy—Margaret Thatcher was."

BIG BIOLOGICS

Today, biologics are dominated by monoclonal antibodies. All monoclonal antibodies are essentially secreted by a single clone of cells, cultured in their millions. Monoclonal antibodies all have the same suffix: *mumab* (or *mAb*, a stylized version of monoclonal [m] antibodies [Ab]). Especially for cancer drugs, mAbs are already blockbusters because an antibody can be modified to make it more hostile toward cancer cells. It can be linked with a payload—another cancer-killing compound. Best of all, antibodies can be made specifically for different cancer antigens, making them a dual- or triple-targeting biological weapon against cancer.

The very first FDA–approved mAb was muromonab, also known as OKT3, for use in preventing rejection of kidney transplants. This was approved in 1986, just two years after Milstein and Kohler were jointly awarded a Nobel Prize in Physiology or Medicine. However, muromonab wasn't a huge success. A significant shortcoming of the molecule was the lack of a suitable myeloma cell line because hybridoma methods were yielding low amounts of antibodies. But better manufacturing techniques and efforts to improve the efficacy were in full force.

I was keen to learn about newer ways to manufacture mAb, and I thought my move to Boston would allow this.

Monoclonal antibodies alone generated a revenue of $15 billion globally in 2008, from there on, to $115 billion market value in 2018—an astonishing amount when you consider the hybridoma method wasn't all that old. In March 2008 when the BTK inhibitor ibrutinib was still being developed, a Danish biotechnology company, Genmab, started a trial with a new monoclonal antibody called HuMax-CD38. This would come to be known as daratumumab—a large molecule that became a major medicine approved in both newly diagnosed multiple myeloma patients and in relapsed patients who do not respond to other treatment options.

I was tasked with developing something better.

Chapter 8

Getting My Feet Wet

*Many have said of Alchemy, that it is for the making of gold and silver.
For me such is not the aim, but to consider only what virtue and power
may lie in medicines.*

—Parcelsus

IT IS ALL ABOUT THE SPACE

December 2018
Greater Boston

In an executive boardroom overlooking the Charles River ten stories below, I
was sitting in a sleek chair listening to a briefing at my new workplace in the
heart of Kendall Square.

Just south of Harvard Square and east of the Central Square neighborhood,
Kendall Square has earned a reputation as a major hub for technology and
innovation and is home to many startups, universities, drug companies, and
research institutions. The neighborhood is constantly evolving, and there is
always something new to see and do here. They say if you are looking for a
place to be a part of something big, Kendall Square is the place for you.

I wanted to be part of something big. Nothing could be bigger than improving cancer outcomes, and for me that was the goal. Here I was learning about
a new mAb for myeloma about to go to clinical trials.

The PowerPoint slide on the screen in our room read "mAb85." The presenter was the vice president of the company, head of drug development, and
more importantly, my new boss. He was a veteran of twenty-five years in the
industry and was looking at his webcam in what appeared to be a library—
through the blurred Zoom background, that much I could make out.

With *mAb* being part of the name of the new molecule, I figured he was talking to the small group of us, in different cities across the United States and Europe, about a monoclonal antibody.

"So as you know," he continued, "the market is crowded. There are established drugs as part of regimens in the frontline [meaning drugs that are given when a patient is first diagnosed with a disease], however, once the patient progresses from the frontline treatments [and they almost always do] there are multiple options for doctors to consider. We need to start there: in later lines of treatment."

He paused to move on to the next slide. This one, a nicely colored, landscape orientation of bar charts showing several marketed drugs along with some code names (meaning they were emerging molecules), was setting the stage for yet another new compound, code name *mAb85*.

I was to lead the development of this molecule.

That said, drug development is quintessential teamwork. While a vice president might head development, make critical decisions, and ultimately bask in the glory of a success (with others, one hopes), they do not control all the moving parts of the project.

Who are these players, you ask, that control the day-to-day affairs then? They were the laboratory scientists who were working on this molecule when I probably didn't even know that I would take up this job; the formulation group who would be producing the molecule as we studied it in humans; my regulatory friends who would communicate with the FDA and similar organizations; and my operations team that would run the clinical trials in some of the most esteemed hospitals in the world. And let's not forget the commercial team, the folks who know the pulse of the market, the value of a brand, and the sales figures. They are the ones who forecast whether a new product, such as mAb85, will have *space* (commercial potential) in the market, and in most instances, how much space that is.

Oh, *the space!*

This is a phrase that is thrown around in drug development by possibly everybody. The meaning of this phrase is highly contextual—as I found out in my initial few days. As a physician, I was concerned with finding new, effective treatments for my patients. I would heal them, advance science, and expand knowledge in the clinical field. Seldom is an academic's work aimed at broad, practical applications.

As a drug developer, my goal was still the same—bringing new medicines to patients—but the stakes were different. Drugmakers must assess the risk in investment before deciding to develop a drug for a disease. That's why questions like the followings are common in drug development unlike in academic settings.

Where are you developing your molecule? (This doesn't mean a high-tech laboratory or your computer. It's the disease area, also called its *indication*.)

Is the space crowded? (That is, Is there a lot of competition?)

How high is the bar in this space? (That is, What is the response of the disease to other drugs?)

Is there space for novel mechanisms? (That is, Are there newer ways a drug could work?)

Do you have a profile for a best-in-class drug in this space? (That is, What are the chances of your molecule becoming the best drug in the category?)

As we saw earlier, when ibrutinib was approved, it was the first-ever drug to inhibit the BTK enzyme so it was a first-in-class BTK drug in the market. *Best-in-class* means the drug has a superior profile compared to other drugs of the same class. There are more nuances when labeling a drug best-in-class.

A lot of it gets down to the molecule itself. What is it that you have in hand? What kind of compound is it and, importantly, what is it targeting? If a drug is novel, it matters less how crowded the space is as the company will consider it a good bet if early data are promising. Speed plays a huge part here. The company might start developing a drug first in an uncrowded space, but others might have a breakthrough to accelerate their own version, and in the end that first drug might not get to patients fast enough, facing a steep competition for being late in the market.

The new molecule that I was concerned with was an mAb against CD38. We know that the CD38 antigen is expressed by all myeloma cells. The mAb attaches to the CD38 antigen, coats the myeloma cells, and invites immune cells nearby to attack them. The antibody-coated cells are also eaten by the specialist cell eaters in our blood known as *monocytes* and *macrophages*. Then there's a cascade of substances known as *complements*, which when activated by CD38–targeting mAb can kill myeloma cells. These antibodies can directly obliterate myeloma cells as well without the help of immune cells.

What's more, CD38 also acts as an enzyme that catalyzes chemical processes outside the cells such as tweaking of the immune cells' messaging. By inhibiting this enzyme function through CD38–targeting antibodies, the constant flow of M-proteins can be stemmed.

All this means CD38 is a great target for cancerous plasma cells.

But a CD38–targeting mAb had already been a hugely successful drug in treating multiple myeloma and approved by the FDA. In my *Games of Thrones* analogy, Johnson & Johnson might have already conquered the iron throne with their mAb, daratumumab.

So what was the plan for me here, with another mAb of the same class?

There were opportunities for improving the anti-myeloma action of CD38–targeting antibodies through enhancing their immune action. Multiple myeloma still poses a huge challenge, and the addition of a better-performing, second-generation CD38 molecule might as well make a difference in our patient's lives.

It all sounded good. I knew the company also had a first-generation CD38 antibody in development. Once the basic platform had been laid, I asked the vice president, "Don't we have a first-generation molecule before this mAb85?"

I was itching to learn if our new molecule could make a splash when compared to something as successful as daratumumab. And how different would the mAb85 be, when compared with its predecessor.

"Yes, mAb85 is the second generation, while we have a first generation in development," he responded.

By then I had been briefed on mAb85's preclinical properties. So I could follow the data on the screen without much trouble.

Before we can treat any human being with a new molecule, we must first define and explain its *pharmacokinetics* (PK) and *pharmacodynamics* (PD) and understand the new molecule's potential toxicity.

Does it bow out of the body soon after it's given (PK)?

Does the molecule hit the right targets? Are there any off-target effects (PD)?

What are the observed toxicities or adverse effects (AEs)?

Answers to these kinds of questions have a consequential effect on the journey of any molecule in its development, all the way to its approval or rejection. If the body gets rid of a new molecule very quickly (short PK exposure), we need to give it more often to the patient. More frequent dosing means additional drug, effort, and cost.

Conversely, if a drug has long PK exposure, it can be given only once and it might exert its effects for the next twelve hours or longer. The downside of long PK exposure is possible unwanted side effects (some side effects are desirable). Along with PK and toxicity, I needed to pay attention to the PD of my new molecule.

Simply put, PD is what a drug does to the body. It is the drug acting on its targets, exerting the effect it was meant to as well as causing any unintended but unavoidable off-target effects. Think of chemical reactions, normal bodily responses, and changes at the molecular level. All of these are happening because the drug is acting on its targets. Together, PD and PK influence how effective a drug might be.

Once these basic parameters are explained and I have seen the preclinical proof-of-concept that the molecule is inhibiting its target, in this case, myeloma cells exhibiting CD38 antigens on their surface, it is time to design a clinical protocol. This document dictates how the investigational drug will be given, how frequently it will be given, and how the patient will be monitored for any side effects. If this isn't done right, an experimental medicine could fail to advance early in its journey.

The protocol, though, is just the start. As a drug developer, I need to envision what this molecule will do in the long run—that is, several years down the road. The entire development path of a new molecule, from first-time-in-human trials all the way to its approval needs to be charted to fully realize its value, and not just its commercial value but how valuable the drug will become for patients. How will this value stack up against the value of other drugs in the same space? In other words, is it a worthwhile investment to develop this new drug? Another key aspect to think about is the new drug's combinability—that is, how well this new molecule will pair with existing drugs. All of this needs to be planned ahead and agreed upon by the key members of the drug's development team, before a large investment is made.

A popular concept thrown around when assessing drug effects is its *dose-response relationship*. This measures a drug's effect when a certain amount of it is given. An efficient drug will achieve its desired effect when an optimal dose is taken then leave the body (be excreted or metabolized) after a certain time, having fulfilled its promise. While we can increase the dose of a drug to make it work better or for a longer period, this runs the risk of higher toxicity or more side effects.

More is not always better—at least in medicine.

At this point, I was getting to know the team, the wheelers and the dealers of the project, but I couldn't help asking a key question about mAb85's expected anticancer action.

"How does the mAb85 respond against myeloma in models compared to the first-generation mAb?" I wanted to understand the presumed efficacy of the new molecule compared to the first generation of the anti–CD38 antibody class. These comparisons are usually done at the pre-clinical phase.

The vice president blinked. Was there a glint in his eyes, perhaps?

"We have preliminary data that indicates the new molecule is several times more potent in attaching to the target than the market leader."

Daratumumab was the leading drug in this space. Nearly 250,000 patients with multiple myeloma had benefited from daratumumab already, and it had raked in $4 billion in annual revenue for its developers! It was a blockbuster medicine.

Was he suggesting the mAb85 could be even better, a potential best-in-class?

As if my question were written on my face, his reply came immediately: "If we can develop this fast, if you can get going with the project ASAP, we have a shot."

The challenge was laid out.

I felt a rush of adrenaline, a strong sense of anticipation for the coming weeks and months. Outside, bright sunshine was melting away piles of snow from a nor'easter of the past week. Across the room, other members of the team stirred in their chairs.

Bring it on!

I had a chance to take down the enemy: multiple myeloma.

WHEN MYELOMA STRIKES

Multiple myeloma strikes at random.

There aren't really any predictors that allow us with 100 percent certainty to say that someone will have it, but there are risk factors for this cancer of the plasma cells. Among them, being a black male increases the risk of having myeloma, but this isn't a modifiable risk: you can't do anything about such a risk factor.

We had started to get a lot of insight from genetic mutations and anomalies at a molecular level related to this cancer, but these genetic changes almost always happened with other cancer-prone mutations. This made treating myeloma even more difficult because the tumors weren't identical between two persons and the responses to standard drugs can vary among individuals.

In their journey to becoming a cancer, malignant plasma cells aren't always uncontrolled bandits from the beginning. Usually there is a symptom-free period when the abnormal proteins are produced inside the body. At this stage, treatment isn't necessary. But rarely, this state of abnormal protein production (called *MGUS*) proceeds to a silent disease (called *smoldering*) and then becomes a symptomatic disease. That's when treatment is necessary as the myeloma cells wreak havoc inside the body.

Pain is how most patients become aware of myeloma, and that's how Jude found out about it as well.

March 2016
Philadelphia, Pennsylvania

After working in the restaurant industry for more than two decades, Jude had decided to go to school. She was training to become a massage therapist and had only two semesters left in her program. She was attending to one of her patients on March 14 when things started to get awry for her.

She felt a sharp pain in her lower back, and it took her several minutes to stand up from sitting. She willed herself through it, but the pain came back the next day. And the following day.

Jude's school wanted her to sort this out. She thought, "I'll get a physical done for the pain and a letter from the doc stating I needed to rest for a few days—that's all."

But the doctor had other suspicions. He referred her to a cancer doctor given her history of having had a lump in her breast, which had been non-cancerous. Jude went to an orthopedic surgeon. She needed to take care of the back pain first; cancer wasn't on her mind.

After the initial x-rays, the ortho doc had a question: "Did you break your back at some point?"

"Not that I know of," Jude replied.

But the imaging showed there was an old fracture in her back. And there was also an irregular break in between her vertebrae that was more recent.

The doctor ordered an MRI. Not fully knowing what was coming at her, Jude nonetheless was preparing to start a new job the following week. She decided to have her MRI over the weekend so she took several painkillers and was training at her job the following Tuesday when a voicemail reached her cellphone. This was the most devastating news one could receive from their doctor.

"Hi Judith, I hate to be the bearer of bad news, but I just got off the phone with your orthopedic surgeon. You have cancer, and it's metastasized." The primary care physician's voice delivered the ominous message.

In a new setting, in front of people she didn't know too well, Jude had a very hard time accepting what she had just heard by voicemail.

She listened to the message again.

The reality of it hit her this time. The news spread in her family like wildfire, and Jude's sister called shortly afterward.

"I didn't know your pain was this bad. I wish you would have told me!" her sister said, sobbing. She drove to Philly the next day to comfort Jude and to take her away from work.

"We are getting a second opinion," Jude's dad declared when she flew home to her parents.

They scheduled an appointment the following week. Upon examination, the oncologist couldn't tell right away what kind of cancer Jude had as her organs weren't involved. But this didn't give her any peace of mind. Didn't the MRI report say the cancer had already spread?

As if to prove the cancer's aggressive nature, her pain became unbearable.

"They took me to the ER, and that's when things got crazy. I was in so much pain and in and out of consciousness. The doctors there talked to my oncologist and said they needed to attach titanium rods to my femurs as a precautionary measure. I had some X-rays, and the surgery happened pretty quickly, maybe a week or two later," Jude recalled.

It turned out Jude's pelvic area was punched with lesions where myeloma had caused her bones to be eaten away. She was under heavy sedation for the extreme pain. After a week in the hospital, Jude was moved to a rehabilitation facility to learn to walk again.

The days became endlessly depressing. It felt so lonely and grim. Jude wanted to refuse everything—the therapy, the rehab, the blood tests—all of it. Her mind started playing tricks; she was seeing things that didn't exist. She was also showing signs of extreme forgetfulness.

That's when they had to rush Jude to the ER for the second time.

Jude's blood became too thick from the collection of copious abnormal monoclonal protein that was being produced by the malignant plasma cells in her body. These abnormal proteins made her blood almost as thick as tar. The thicker a fluid is, the less it flows, and in Jude's case her blood wasn't flowing properly to her vital organs: heart, kidneys, brain. This was resulting in her fogginess due to a lack of blood supply to her brain. This was a serious condition. She needed an urgent remedy, known as *plasmapheresis*.

Plasma is the liquid part of blood. In plasmapheresis, doctors exchange the thick blood for fresh plasma that does not have abnormal monoclonal proteins. It can relieve the symptoms of cancers such as myeloma and provide temporary relief but it won't cure them. The uncontrolled plasma cells will pollute the exchanged plasma to make it as thick as before.

It was only a matter of time before things would get frantic again.

Jude needed treatment for her myeloma. It was clear that she had stage 3 cancer, and with such an advanced disease, immediate treatment with a cocktail of drugs was the usual treatment.

Jude received a combo of several drugs, some of which were already standard treatment at her stage of the disease such as velcade and steroids. Along with the standard drugs, she received daratumumab, which targeted CD38, once a week for twelve weeks. The oncologist gave an educational brochure regarding the medicine to her family to explain how an mAb works.

"It wasn't a true chemo. It was monoclonal antibody therapy. I didn't lose my hair with this," Jude said.

Indeed, biologic drugs such as daratumumab are a testament to what modern, targeted medicines can do: improve the patient's depth of response in the first phase of treatment with fewer side effects. With daratumumab in the arsenal, Jude's myeloma was sufficiently controlled so she could move on to the next phase of her treatment: stem-cell transplant.

The stem cells are special cells that can turn into many kinds of cells in the human body. They can make copies of themselves, self-renew, and give rise to other cells such as blood cells, muscle cells, or brain cells. Inside the hollow of our long bones, the blood cells are formed. In cancers like myeloma, the bone marrow is invaded by cancerous plasma cells. If these plasma cells are killed with chemo to make room and then the bone marrow is planted with new stem cells, it can start forming healthy blood cells again.

For the transplant, Jude was admitted to the hospital for the doctors to collect stem cells from her. Fortunately she had enough active stem cells for three transplants.

She enjoyed the stem-cell transplant process.

"When they give you the stem cells, it's really pretty cool because you can see them going down the IV." Jude described her fascination at observing the cells passing through the IV tube attached to her.

The second stem-cell transplant took place about six months after the first. Once that was done, it was time for a bone marrow biopsy, a test that gives the verdict on how the treatment is faring against the myeloma.

All of it—daratumumab, chemo, surgery, conditioning, stem-cell harvesting and transplant—led to a magical result: "You are unquestionably in complete remission of your cancer!" said the doctor upon reviewing the biopsy report.

Motionless, Jude just sat there.

How do you react to such news? The pain, the suffering, the near-death events, the loneliness, the depression, the countless hours of being a patient, the feeling of not knowing what to plan for—all these memories came rushing at her and she just didn't know how to react.

She walked out of the hospital by herself, in complete remission of multiple myeloma.

When I was working in my office in Kendall Square, about 135 miles away in Boston on a frosty day in December 2018 planning to develop better and safer drugs for myeloma, Jude finished her school program and became a licensed practitioner.

Today Jude practices as a massage therapist. Once in a while though she looks at her rear-view mirror, trying to pick up any sign of a pursuit from that tyrannical emperor of all maladies. Multiple myeloma remains an incurable disease. In a year about 35,000 new cases are diagnosed in the United States and more than 12,000 pass away due to this cancer.

THE AGE OF TRIPLETS AND QUADRUPLETS

In my discussions with the experts, it was obvious that a CD38–targeting mAb had become an important drug in myeloma. Patients were living longer with the disease, and new drugs (not just daratumumab) were becoming more accessible. The challenge now was to figure out what was the best cocktail, or combination of drugs, to give to patients earlier in the diagnosis.

As an initial treatment, a triplet of drugs, including velcade and revlimid (VR) along with dexamethasone had been shown to be superior to a doublet of just revlimid and dexamethasone, making it nearly 30 percent more likely patients would live longer and disease-free. My colleague, myeloma expert Nina Shah puts it this way: the triplet is a "Honda Civic"—it's available, trusted, and it does the job well. Doctors needed to first make sure that triplet use was feasible for patients. This data was presented in scientific meetings, including at the ASH annual meeting in 2015, and the data supported triplet use.

So triplets were better than doublets.

Soon quadruplets entered the picture. Now we had comparative data between a triplet regimen and a four-drug cocktail (the addition of daratumumab to the original triplet, meaning VR plus dexamethasone plus daratumumab). And joining the field a bit later, isatuximab cocktails also indicated the value of a quadruplet regimen.

As more drugs were added to the treatment, the chances of side effects increased. On the other hand, the benefits were undeniable.

So do the experts pick a triplet or a quadruplet cocktail?

A key piece of the puzzle that heavily influenced the decision was to find out if a patient with myeloma could get a stem-cell transplant. For many oncologists, the possibility of a stem-cell transplant should be weighed first, and whatever first treatment got the patient to that transplant got the preference.

Usually a triplet cocktail is used to get good control of the myeloma and to prepare the patient for a stem-cell transplant.

But the quads were pushing ahead with more data, and the findings were becoming more convincing each time the results are presented. For now, quadruplets are an option for our patients, and in the not-so-distant future we would know which patients benefitted more from a quadruplet.

Meanwhile, myeloma still found ways to escape the onslaught of the triplets, quadruplets, and stem-cell transplants. The malignant plasma cells managed to evade the actions of the newer therapeutic arsenal—including CD38–targeting mAb.

We need to keep attacking this enemy with novel immunotherapies and reduce the burden before it became an advanced cancer.

I planned to try to do just that.

Chapter 9

The Challenges

If you're not making mistakes, you're not taking risks, and that means you're not going anywhere. The key is to make mistakes faster than the competition, so you have more changes to learn and win.

—John W. Holt Jr.

KICKOFF

January 2019
Greater Boston

I was studying the path of daratumumab's development very closely.

One key aspect of working in the biotech industry is paying full attention to the development path of your competitor. In my case, within the CD38–targeting class of drugs that were mAbs, mAb85 was going to be the second generation. Of course the onus was on us to create a better medicine, to show patients, physicians, and, most importantly, regulators that the new drug had a superior safety and benefit profile.

We wanted to show that mAb85 could be the best-in-class CD38 antibody.

Even with a phenomenal and theoretically better drug, however, becoming best-in-class is much harder than becoming first-in-class. In theory, the cleanest way for me to prove mAb85's superiority would be to start a clinical trial in multiple myeloma patients (at the same stage of the disease and with similar features and inclusion criteria) then treat one group with mAb85 and the other group with daratumumab, and if the mAb85 group showed better results, this would prove it was a better drug. Sounds simple, right?

Except it doesn't work like that.

For starters, regulators wouldn't allow me—or any drug developer—to treat advanced-stage myeloma patients with an unproven molecule if a tested-and-proven medicine like daratumumab was available. First, the new drug must go through early clinical trials among patients who don't have options left. In these situations, the patients will have received pretty much all approved treatments but their cancer will have progressed. These patients have what is known as *relapsed-refractory* (R/R) disease with no standard of care (SOC) left for them.

That's where a new molecule like mAb85 begins its journey. In our case, multiple myeloma patients who had received anywhere between three to six prior treatments and had progressed would be allowed to receive mAb85 so we could study its effectiveness. If the molecule, administered all on its own, showed benefits at this stage despite the patients' previous poor responses to similar medicines, then we have a drug on our hands. It's only at that point that we could take it up with regulators to propose a possible comparative study between mAb85 and an approved medicine like daratumumab—if we wanted to. A lot of other considerations need to be addressed before a drug company goes to a direct, head-to-head comparison study.

But first we'd need to prove efficacy among progressed myeloma patients. We were not there yet.

The one major benefit of mAb85 being a later molecule within the same class of drugs compared to daratumumab as a groundbreaking, first-in-class drug was that I could study daratumumab as a forerunner medicine. Because daratumumab was already approved and had been available in the clinic for a few years, there was publicly available data on it. The investigator physicians who oversee drug company trials publish these findings on each drug's benefits (patient response rates, survival benefit), its toxicity or side effects, and any appropriate warnings. This is how the community gets wind of new and exciting treatments. The FDA receives these data packages and reviews them thoroughly to consider a drug's approval.

I had followed daratumumab's journey from the other side of the aisle (as a physician-scientist). I knew that daratumumab, just like most drugs, received its first approval in patients with multiple myeloma whose disease was R/R. The first of the daratumumab trials were GEN-501, which started in 2008 and completed enrollment by early 2014. This trial was conducted by Genmab, the company that created the drug and codeveloped it with Johnson & Johnson.

The second key trial. called SIRIUS started after GEN-501 had nearly completed its journey. Between GEN-501 and SIRIUS, 148 myeloma patients received daratumumab as a single drug after having progressed from about five previous treatments. Overall, about 30 percent of the patients responded to daratumumab, and they remained in good control of their myeloma for eight months or more. The patients survived for up to 28 months—a significant improvement when you consider these patients did not have other options of treatment.

My fellow physician-scientist and myeloma expert Saad Usmani had been involved since the early days of daratumumab development. Although I was deeply involved in the ibrutinib saga at the time GEN-501 completed enrollment, the promising data from the GEN-501 trial was discussed in our group in Houston.

Now I was interested in revisiting and studying these early trials of daratumumab. My new molecule would be going into a clinical trial with similar late-stage myeloma patients. There were subtle differences in the way mAb85 fights myeloma compared with daratumumab. But since the drug target was the same as daratumumab—CD38—I could try to predict what kind of side effects or safety challenges I might face in the first trial.

In first-in-human trials we are specifically concerned with toxicity because these drugs have never been tested in humans and you don't quite know what surprises they might present with. As we were full steam ahead to complete the clinical protocol with mAb85 that would direct investigator physicians on how to administer our drug, I decided to dig deep into daratumumab's early clinical experience.

IT TAKES ONE PATIENT

Vigor and a willingness to learn are seen in top scientists and doctors. An example of this was Wang, Rule, and Sharman in their quest to bring small-molecule-targeted treatment to lymphoma and leukemia. For myeloma, I had learned a great deal from Saad Usmani. It was especially relevant for me to hear about his early clinical experience with daratumumab.

Down in Southeast Asia, by the shores of the Indian Ocean, lay a land prone to calamities. Powerful hurricanes battered its coastal areas regularly. Monsoon rains caused its mighty rivers to swell and flood the plains for months on end. Poverty ravaged its communities, and in 1971, a bloody war was raging between its powerful neighbors.

My native land, Bangladesh, was burning.

It was fighting for its freedom when India joined forces with the local militia. By the end of 1971 after a bloody war that lasted for nine months, Bangladesh had achieved its place on the global map as an independent country, liberated from the oppression of Pakistan's armed forces.

Usmani's father fled the war in Bangladesh. He moved to Karachi, the largest city in Pakistan, amid the armed conflict in the region.

The 1980s were an interesting time geopolitically and especially for the Indian subcontinent. Ever so keen on meddling in politics, the Pakistan Army and its general Zia-ul-Huq imposed martial law on major cities in Pakistan. Islamization of the country was in full swing, and Karachi stood as one of the few liberal places in all of Pakistan. There were tensions and uncertainty in the region, especially as the Soviets effectively took control of neighboring Afghanistan. Pakistan had a huge foreign debt then just like it has now. Energy costs were crazy.

The Cold War was on.

Young Usmani did not care much though. By then he had traveled to many places with his parents and was having a good time. Usmani's dad worked in Saudi Arabia, Islamabad, and Lahore, studied in Milwaukee in the United States, and then went back to Lahore.

"The longest period I lived in a place is possibly Lahore, after North Carolina!" Usmani said. He had fond memories of his early travels.

Back in Lahore in 1996, after his dad had completed his studies in Milwaukee, two things happened to Usmani. First, his love-affair with the game of cricket faced a crisis, and second, he got a chance to study at a prestigious medical school in Pakistan.

Usmani was a cricketer at heart. The game of cricket has a crazy following on the Indian subcontinent. Pakistan happened to win the premiere cricketing event—the World Cup—in 1992. Usmani got called up for a regional junior cricket team in Lahore, which he thought was his destiny. He was a hard-hitting batsman and an occasional bowler too, an all-rounder cricketer, as they called the players who could both bat and bowl.

His father had other ideas.

In hindsight, Usmani didn't have a chance, really. Becoming a medical student is a super-competitive process in Pakistan. Each year aspiring youngsters take part in the entrance exam nationally and only a talented few can get in. There was no way he was going to leave medical school to become a full-time cricket player.

He completed medical school in Pakistan and moved to the United States for further studies and training.

Let's fast forward to the first decade of the new millennium. In 2007, Usmani was training at the University of Connecticut at a time when the entire cancer research field was buzzing with new targeted treatment approaches. As a young fellow, Usmani longed for hands-on laboratory research experience as he knew he needed this to understand the action of drugs. At the University of Connecticut at the time there weren't many opportunities for that. But he managed to get in touch with one of the laboratory scientists at the institution whose work was focused on a specific kind of protein in humans known as a heat-shock protein (HSP).

In our body, cells are always bathing in a changing environment. Despite all the chaos of the cell environment, the quality and balance of protein production in the body must be preserved to maintain the body's integrity. Proteins are essential to carry out cellular functions, and they must have correct structure and balance.

If proteins are malformed (misfolded), this can lead to a variety of diseases. Fortunately we have a large family of proteins in our body that act as protein-police: they are in charge of making sure that newly formed protein chains (known as *polypeptides*) are correctly folded. They are tasked with directing the formation of proteins and destroying them if they are misfolded or separating them when the proteins are clumped together. HSPs make up about 5 to 10 percent of the total cell proteins.

Occasionally the HSP production goes haywire. Sensing stress (even though it may not be there), abundant HSPs are produced, which leads to multiple diseases. There are several classes of HSP, all identified by molecular weight. For example, HSP90 has a molecular weight of 90 kilo Dalton. An overabundance of HSP90 has a role in multiple cancers including leukemia, breast cancer, prostate cancer, and lung cancer.

Usmani got his hands on an HSP90 inhibitor molecule. That's when his focus shifted to myeloma, a cancer resulting from high amounts of monoclonal, abnormal protein. He teamed up with Gabriela Chiosis, a medicinal chemist at Memorial Sloan Kettering in New York, to study HSP90 inhibitors in multiple myeloma. Usmani studied myeloma cells under the microscope in the presence of HSP90 inhibitor. What he saw hinted at myeloma cells being suppressed by the inhibitor, so he decided to present his findings at an ASH annual meeting. His training was coming to an end soon. Usmani wanted a new job.

During the poster session at the meeting, myeloma pioneer Bart Barlogie stopped by Usmani's poster board.

"You should come and work in Arkansas," said Barlogie, as he reviewed the poster board. He had founded the Institute for Research and Therapy at the University of Arkansas for Medical Sciences (UAMS) and was a champion clinical researcher there in myeloma.

Usmani was thrilled.

He ended up taking the position and moved to Arkansas from the Northeast.

"There is no other place that can replicate the level of excellence and research you can do in myeloma in Arkansas," he said. Usmani was amazed at the quality of the facility. "Everyone who's been to the center accepts this as a fact, because Bart is an exceptional scientist with all those resources!"

Usmani worked at UAMS for three years. During this period he had access to a wide variety of new, experimental medicines. We were in a biotech boom, post genome sequencing era when a lot of small molecules, antibodies, were starting to make their way to preclinical laboratories as their creators hoped to make it to a clinic. He had a great chance to interact with small and big biotech and pharma companies and explore new ideas for clinical trials.

I was aware of a drug called carfilzomib that Usmani tested during this period. I knew from the published research that daratumumab and carfilzomib had some interesting commonalities. While carfilzomib is an entirely different class of drug—a *proteasome* (a small cellular complex that breaks down proteins) inhibitor, I wanted to learn more from Usmani about the interesting commonalities between carfilzomib and daratumumab.

Since daratumumab and mAb85 both targeted cells expressing CD38, it was not hard to see how patient reactions might be similar. The more I could understand, the more I could anticipate problems and try to design mAb85's administration protocol around them. Early, successful administration would save time.

Carfilzomib was given as a compassionate-use treatment protocol UAMS had at the time. If a patient has a serious, immediately life-threatening condition, they can receive an investigational medical product even outside clinical trials when no alternative option is available. Carfilzomib wasn't approved for myeloma at the time, but through this pathway, some patients received treatment with it.

"So in Arkansas, we could give one cycle of carfilzomib and if there is no response to the drug, we could add other therapies for myeloma," Usmani explained to me.

The drug was originally given as an IV injection between two to ten minutes. Because of this range, the nurses tended to give the injection in two minutes. The first twenty patients who received the drug started having difficulty breathing, coughing, pain in the chest, and pneumonia.

Evidently something about the drug was creating cardiopulmonary problems.

"Intuitively I knew something was off about the way the drug was given, leading to these issues, then I had an idea!" Usmani's face lit up as he was narrating the events. "I spoke to Bart."

Barlogie agreed Usmani's idea made sense.

And so Usmani directed the nurses to give the infusion over a period of thirty minutes—that is, very slowly. The results were impressive: a dramatic decrease in cardiopulmonary problems in the patients, all of whom used to have them before. The slower infusions made a difference.

Possibly carfilzomib got a second life as a medicine due to the slower infusions Usmani advocated. When the drug was approved by the FDA in 2012 for the first time in the treatment of R/R myeloma, the approved label indicated cardiopulmonary warnings. No one was too concerned—the "split dose" concept negated the immediate reactions of carfilzomib.

By the end of 2011, early data were starting to appear in conferences from the single agent GEN-501 trial on daratumumab. About two-thirds of the patients in the GEN-501 Phase 1 trial, which had started in 2008 when Usmani had just joined in Arkansas, experienced what was known as infusion-related reactions: breathlessness, coughing, and respiratory infections.

To combat these infusion-related reactions, similar to what Usmani did, physicians decided to do two things: first, the patients received only 10 percent of the full dose before they could receive the standard dose; and second, the patients received the infusion over a longer period of time. The first infusion of daratumumab lasted for about eight hours and the second infusion about seven hours. Later these infusion times were reduced quite significantly.

This was an awfully long time for a medicine administration. But for Usmani and other myeloma experts who advocated such measures, this was the right way to give the drug intravenously. It controlled the rate of infusion-related reactions that were occurring commonly before.

In the summer of 2013, Usmani moved from Arkansas to North Carolina. At that time, buoyed by good responses to daratumumab in the GEN-501 trial, Johnson & Johnson started the SIRIUS trial. There was a lot of excitement in the myeloma community about this. Usmani was eager to enroll patients in the trial, but the Johnson & Johnson liaison told him that in order to have any chance of enrolling even one patient he needed to open his clinical site within two months.

So he did.

"But was there a moment when things weren't looking good?" I asked him. Deep down, I was thinking of a possible setback of my own. Surely not everything was rosy for daratumumab—it's hardly ever the case in drug development.

"Yeah, it was similar to your BTK story."

Usmani recalled that a patient had a sudden surge of lymphocytes count in their blood just as Sharman and Rule experienced with ibrutinib.

"I thought, s—t, maybe, you know, I am giving a second cancer or something serious to this patient." Usmani was genuinely concerned.

This patient had an 800 percent jump in their lymphocytes after they received the first dose of daratumumab. That kind of rise in the count is almost unheard of. But as it turned out, there was an expansion of a particular kind of cell, T-lymphocytes, in this patient. These types of T-cells are also known as *killer T-cells* because they kill cancer cells, cells infected by bacteria or viruses, and damaged cells. So although the high number of killer T-cells was unexpected, and frightening for the monitoring physicians, it wasn't a bad thing for the patient: the boosted T-cells were the enemy of cancer.

Usmani shared his findings of lymphocyte counts with his peers. He also reached out to the Johnson & Johnson team, who started looking at other patients' data from the earlier GEN-501 trial. In the end, this very first patient provided a hint that in addition to killing cancer cells directly, daratumumab also bolstered natural immune cells—kind of a second wall of defense against myeloma. What's more, this boosted immune effects readied the body to respond better to other drugs. This was a boon for Johnson & Johnson, who went on to start other trials that combined daratumumab with multiple anti-cancer medicines.

Today, combination daratumumab regimens are a powerhouse in treating newly diagnosed patients with multiple myeloma, and the vast majority are treated pretty much chemo-free.

So sometimes it takes only one patient to form your perspective on a new drug—whether it will make it or not. For Usmani, there was one patient who convinced him the drug had a future.

"This was a very refractory patient," he said. "Received two other classes of drugs and stem-cell transplant, never went into a complete response." These patients are considered high risk.

"Yet within three cycles of daratumumab, they were in complete remission!" Usmani knew then and there this mAb would be big.

DIVIDE AND CONQUER

Back in our camp in Boston, we knew what we needed to do with mAb85.

For the first-in-human trial, we proposed to administer the first dose in two parts: a part of the full dose of mAb85 would be given on the first day and the rest on the following day. To anticipate and manage any reactions, the infusions were to be given slowly and the patients would be monitored at the hospital during the infusion, in case they needed support.

As we approached the day when the first-ever patient would be treated with mAb85, I had a sense of restlessness. What if they had an allergic reaction to the new drug? What if they got very sick? How about not being able to tolerate the treatment?

In the laboratory plates, our second-generation antibody was performing better than the first-generation CD38 antibodies. This drug was more agile in binding to its target and had enhanced attractions for activating immune action. When it came to treating myeloma cells, the drug showed better cell-killing efficiency in laboratory experiments. All these pointed to a strong molecule with a potential for causing infusion reactions.

I thought about the first patient as we were about to treat. The patient's labs looked good and met all the eligibility criteria, and thankfully, their myeloma wasn't exploding. The very first patient in a Phase 1 trial is always an anxious moment. The patient, the treating doctor, and the developer of the drug, everyone has every hope the drug will be off to a good start.

Hope—a four-letter word we sometimes use casually in daily conversation. But if you are suffering from an incurable cancer (I hope you never will) and have exhausted all treatment options, hope is your best friend. Hope is the light that guides you through the darkness of depression, the fire that keeps you warm in the cold, and the anchor that keeps you grounded when everything else is spinning out of control. You have to believe that things will get better even when clear signs may not be available yet. Hope is the strength that keeps us going when we want to give up.

Think of yourself as a patient who is about to start an experimental treatment for an aggressive cancer. You signed up for a first-in-human trial because there likely isn't any proven medicine that the doctor can give to you anymore. But you so dearly want to attend a graduation, maybe your son's or daughter's, or perhaps it's your own graduation. Or you long to spend loving moments with your partner on a crisp October morning on the porch, basking in the soft glow of the sun when the leaves are turning shades of yellow. How about that long-awaited skiing trip on the mountain slopes when the snow has fallen? Or swimming in that pristine lake now that spring is in the air? The feel of the golden soft sands under your feet on a beach in the summertime, holding hands with your lover—this can't just go away. How can you let go?

All this rests on the hope that there is a remedy.

The patient sits, likely with a close family member or maybe alone because that's what they prefer. The doctor and the nursing team then go through the informed consent—a document that explains that the treatment about to be given is experimental. The informed consent lists all the risks that could be associated with the experimental molecule and the possibility of serious, sometimes deadly complications that could happen. The possible benefit of an experimental drug is largely unknown especially in an early trial.

The patient signs this consent. He or she is desperate for a remedy.

The lab works are typically complete about a week ahead. There are some tests that are done on the day of the first treatment a few hours before. The patient may also receive a cocktail of medicines that prepares the body for the investigational drug infusion along with any other routine medicines that they take.

When the moment comes, the nurse will attach the IV tubes to the drug bag and start the infusion.

You are possibly looking at the infusion drip, tiny drops trickling into the lines, carrying the molecule to your bloodstream where they will fight the cancer cells. Like Jude, you might be fascinated by the process. Like Bobby and so many others, you might get a lasting good response from this new medicine or cocktail.

Conversely, your cancer might not respond to the drug at all or you could get a not-so-good response. Maybe there are more side effects than expected. No one really knows what's going to happen for sure—that's why it is called a clinical trial. Without trials it is impossible to measure the benefit versus the risk of a drug. Tomorrow's cures are dependent on today's clinical trials. Hope is great, but it must be based upon a solid foundation of groundbreaking research. Without our patients and caregivers, we won't be able to do such research.

A GOOD START

In a global clinical trial, patients can give their consent and be enrolled though they live on an entirely different continent than the project team because different clinical sites can initiate their participation at different times and in different places. The hospital where the first patient received mAb85 was located hundreds of miles away from Boston. It was off-hours for us when the infusion started because of our different time zones. But I asked our local team to update the global team members as soon as they could.

I was impatient.

On that July morning, rushing through the commute in Kendall Square, I was at work before 7:30 a.m.

The email was waiting in my inbox.

The patient had received the first part of the dose of mAb85 uneventfully. There was a mild cough for a few minutes or so, but nothing serious at all. They completed the infusion and were resting comfortably. Eaten well and slept well. Nothing unusual in the bloodwork.

We were off to a great start. Later in the afternoon, I called the oncologist to thank her. She was an experienced myeloma expert and a caring person. Wise and warm, she was a great teacher and a scientist. I asked her to enroll more patients if she could.

The success of a new clinical trial, especially in Phase 1, depends on the drugmaker's ability to encourage medical practitioners in various cities and countries to enroll patients. Drug development is super competitive, and there are only a handful of cancer centers reputable enough to recruit patients to a trial. There could be competing trials ongoing in these centers so a good relationship with a center's doctors and research coordinators was essential. My goal for mAb85 was to engage with leading oncologists as much as I could, providing them with timely updates and kind reminders of the need to screen for eligible patients.

We needed to go through the first few cohorts soon. Like the first patient treated that morning, these patients would receive a low dose of the drug to monitor for safety concerns. But it would only be after this period of safety observation—known as a *dose-limiting toxicity* (DLT) period and typically about a month—was over that we could move to higher dose levels that would be more likely to be of therapeutic benefit. For mAb85, the prespecified DLT period was twenty-eight days—typical for this class of drugs. When treating aggressive cancers, you can't afford to stay at a low dose any longer than necessary.

The oncologist promised to offer the trial to a few eligible patients.

By September we had enrolled the first cohort of patients. The early cohorts of a novel molecule are the most challenging. Physicians know we start at a small dose, and not surprisingly they don't want to put their patients on a likely sub-therapeutic dose. But starting low is the safest way to ensure we are not harming the patient with a high dose by causing unwanted toxicity. The regulators are very strict and clear about this and rightly so, but it isn't always the most hopeful option for the patient. Usually, until you reach the second or third dose level, the likelihood of a quick anti-cancer response is low. But there are many instances where even the first dose in a clinical trial has resulted in a remission of the cancer. We'd seen that with ibrutinib and daratumumab and other targeted treatments.

So I was relieved that we didn't have any serious toxicity in the first cohort.

It would have been very difficult if the first few patients had experienced a lot of side effects and were not able to complete the infusion. That didn't happen. We did not have a DLT that would have put the brakes on going higher on the dose. The only question was the milder infusion reactions, and these were managed very well by the experienced oncologists in charge.

But the second patient had a curious event that confused even her veteran physician.

"She had a small patch—about one and a half inches in diameter—of discoloration on the chest—a minor bruise," the oncologist told me.

"No bleeding?" I asked.

"No active bleeding or anything. It's ecchymoses, so under the skin. We decided to record it as a Grade 1 adverse event—just so you know."

Grade 1 events, the lowest graded side effects, are very common and usually don't cause big issues. A week later the patient had reported that the skin bruising was gone within seventy-two hours. I didn't think much about it then. To me it was a minor event of little significance. My perception changed during the next few months.

Global clinical trial operations are a complex business. There are hundreds of variables—patient factors, country and site level issues, drug supply, the role of the local coordinators, to name a few—that play strong parts in a trial's success. Above all, a stable geopolitical system is important when you're conducting clinical trials. But like cancer cells that lie undetected beneath the surface, something ominous can lay in wait before suddenly upending everything. Nobody was prepared for anything like it.

ENEMY UNSEEN

Somewhere in the Far East

The bat lived in a cave, a place so dark and damp perhaps the only light that reached the cavern was through a few small cracks and holes high up on its walls. The air was thick with the smell of mildew and decay. The only sound was the dripping of water from the ceiling.

The cave belonged to *Rhinolophus* colonies. The bat hung upside down and it fed on insects. This was its den, its own corner. *Rhinolophus* carried many viruses and bacteria in its body. It was a reservoir, a natural habitat for these tiny bugs living and growing inside their mammal host.

This *Rhinolophus* had a virus in its system that wasn't supposed to trouble the bat or even other animals unless the virus found a way to jump from the bat to another species, a process known as *zoonoses*.

That's exactly what happened.

The virus needed the help of another host other than the bat. Perhaps it was a raccoon dog. Or maybe a civet cat or badger—it's very difficult to say with certainty. But whatever the intermediate host was, it inevitably came in contact with *Homo sapiens*—the most evolved species on planet Earth.

And the virus was transmitted.

The human system hadn't experienced this virus before. It was SARS-CoV-2, a variant of coronavirus.

December 2019
Wuhan, China

The sprawling capital of central China's Hubei province, Wuhan, is a bustling commercial center with a population of just over 11 million. In downtown Wuhan, and close to a major railway hub of the country known as Hankou, is the Huanan seafood market. The market is open every day and is a popular spot for both locals and visitors. With about seven hundred stalls in very close proximity to each other in an area roughly the size of a soccer field, Huanan market is separated into two zones, eastern and western. The western zone mainly sells live animals and seafood while the eastern part does its business in livestock meat.

It was business as usual at the Huanan market. In December 2019, several vendors in the western zone sold live wildlife and animal products to interested buyers, anything from snakes, pheasants, and sika deer to badgers, hedgehogs, and giant salamanders. Together with the extremely crowded conditions, the carcasses of the butchered animals and the stray animals and their feces made this so-called wet market a prime location for virus transmission.

From animal to man, and from man to man.

No one really knows who got infected with the new virus first or if indeed the first human to be infected with this virus was linked to the wildlife trade at Huanan, but on December 10, a vendor at the market fell ill with a mysterious, pneumonia-like illness. Soon after, an elderly couple presented to Hubei Provincial Hospital with what presented as large, ground-glass opacities in their lungs that were distinct from the other cases of viral pneumonia treated by the doctors in the area. The couple apparently had no connection to the Huanan market. Their only son who accompanied them did not have any symptoms but had similar unusual findings in his lungs. This family of three was the earliest-known cluster of infections with the new illness.

By the end of December, at least four other workers from Huanan market had been admitted to the regional Hubei hospital with an unknown respiratory disease. At the same time, Wuhan Central Hospital started to receive patients with this disease too—both with and without an apparent connection to the wet market. Wuhan and Hubei Centers for Disease Control (CDC) were on high alert by New Year's Eve.

A major outbreak was brewing.

December 2019
Orlando, Florida

Early December was an exciting time for blood cancer experts, clinical researchers, and drug developers alike. We had waited all year long to come together at our premiere event, the ASH annual meeting and exposition. For the meeting's sixty-first iteration, we were headed to Orlando—the City Beautiful.

Oblivious to what was unfolding in Wuhan, our team was very energized. Earlier in June at the American Society of Clinical Oncology's annual meeting in Chicago, our team had presented emerging results from our first-generation CD38 mAb isatuximab study. This was a randomized Phase 3 study, meaning a group of patients with R/R multiple myeloma had received an isatuximab combo while the other group received the standard combo only without isatuximab. Among the three hundred or so patients treated by June

2019, about half of them had been randomly chosen to receive isatuximab. The group who received isatuximab had been free from progression of myeloma for almost a year while the other group had been progression-free for only about six months.

While the survival data for isatuximab wasn't mature enough to show statistical significance at the time, clearly the addition of our first-generation CD38 antibody was benefiting the patients with previously treated myeloma. What's more, the isatuximab group appeared to have a better survival rate than the non-isatuximab group—a measure that trumped everything else.

Based on this, the FDA had notified us that they would be reviewing these results for a potential introduction of isatuximab in the market under a biological license application (BLA). As a drug developer, you can't possibly be in a better position than when one of your molecules is headlining with strong data, possibly leading to an approval! The FDA had confirmed a target date of April of 2020, by which time they expected to reach a decision on isatuximab approval.

Of course it was too soon to present any data from the mAb85 trial, but we had several sessions in the conference where investigators planned to present findings from the ongoing Phase 3 trial of isatuximab. Expectations were high in our camp as the data readouts looked great.

With that positive feeling, I mingled with colleagues from either side of the aisle—academia and the industry—inside the immense halls of the Orange County Convention Center in Orlando. Between the presentations, we hosted meetings for doctors and scientists. We were trying to spread the word about our second generation of CD38–targeting mAbs. I caught up with several blood cancer experts in the United States and in Europe to update them on our program and to learn about new treatments.

Michael Wang and I chatted briefly outside a ballroom.

"Looks like they are treating you well, Mak!" This was the first time we'd seen each other since I left Houston.

I smiled and congratulated Wang on the new findings for MCL—the cell treatments, a radically different approach from antibodies, were producing some extraordinary responses in lymphoma patients. He asked about my current projects, and hearing about my venture in the CD38 space, he said, "Your CD38 antibodies don't work in lymphoma—I tried that!"

Wang was correct. The CD38–targeting mAbs were in trials for several lymphomas and didn't look like a promising drug there. It was yet another proof that even within blood cancers, one drug isn't fit for all types of the disease. CD38 might be an excellent target for myeloma and not so much for other types of blood cancers. In the war against cancer, we need to replenish our arsenal continually: new targets, new drugs, new ways to treat.

No one in Orlando talked about a new virus spreading in China. To be fair, nobody really knew about it in December 2019. Even if some of us knew, a regional outbreak of an infectious respiratory illness on the other side of the world would not have caused a lot of stir in our minds.

THE PLATELET CONUNDRUM

As the world celebrated the new year still largely unaware of the rapidly spreading virus in China, the Huanan seafood market was shut down by Chinese authorities on the first day of 2020. They notified the World Health Organization (WHO) about a new viral illness.

I was pushing hard to continue enrolling mAb85's Phase 1 trial. The enrollment was slow but steady. Every Phase 1 trial's main objective is to establish a dose of the drug that will be safe and is expected to control the disease. This dose is called the Recommended Phase 2 Dose (RP2D). As the name implies, the idea is to find an RP2D in Phase 1 trials and then start Phase 2 or other advanced trials with patients receiving that RP2D.

Based upon the prior experience of our team, barring any major issues, we were nearing an RP2D in the Phase 1 mAb85 trial that was considered sufficient to treat myeloma. It was an exciting moment for me as I felt physicians would feel comfortable enrolling patients from then on. I mentioned several times in our weekly meeting with all participating doctors and their staff across the globe of the likelihood of an anti-cancer response to the current dose of our drug in trial.

The path to finding an RP2D can be complex. We needed to review all doses and each side effect that the patients encountered at all dose levels; the patients' laboratory reports; and of course the drug's pharmacokinetics (PK) and pharmacodynamics (PD) and their relationship to dose and clinical benefit. From ibrutinib and daratumumab experiences, I knew that we could expect to see a sudden rise in lymphocyte counts in our trial patients although I also knew the reasons for this might be different for our drug compared to the reasons for the other two drugs.

But something different happened with mAb85. The lymphocytes remained fine, but upon reviewing the labs of the patients, I noticed a confusing trend: a decrease in one of the blood cell counts, that of the platelets.

Platelets, also known as *thrombocytes*, are charged with the role of stopping bleeding or preventing bleeding from happening in the first place. The same platelets also become activated via several biochemical drivers in response to injury to a blood vessel. The activated platelets take up a star-shaped form, and together with the help of chemical messengers form plugs to block damaged blood vessels—known as *hemostatic plugs*. While useful for stopping a leaking blood vessel, continued platelet activation and clumping contributes to heart disease and strokes. A fine balance between resting and activated platelets is needed for normal functioning. If platelet counts become too low, the risk of bleeding is more. If the count is too high, well, that's also deadly due to the risk of strokes.

Low platelet count is a common complication in cancer patients especially when they receive chemo. How critically low the counts are going to be, or how long the platelets will remain at a smaller number, depends on the nature of the chemo as well. If the chemo agents are intense and given at a high dose to make room for stem-cell transplants as Jude had before her transplant, platelet counts can get low because the chemo kills cells in the bone marrow. On the other hand, reduced-intensity chemo might not cause a sharp decline in platelets.

But mAb85 wasn't a chemo, it was an mAb.

Still we observed a decrease in platelets—*thrombocytopenia*—on the second day of treatment in some of our patients. I didn't think it was an issue. Curiously, the platelets had returned to normal levels on their own by the end of the first week. The most important part in all this was that there were no active bleeding events or hemorrhage so clinically this was not a problem for the patients. None of the oncologists who treated these patients thought it was a problem either.

But we needed to understand it more. In a clinical trial, every single unusual finding, however trivial, needs to be followed up closely. I would rather pick up an issue early in a small number of patients than ignore it for the time being only to be confronted by the same issue later in a broader cohort.

I remembered the second patient in the trial who had an unusual, minor bruising under her skin that disappeared on its own after seventy-two hours. Now with laboratory data in hand from more patients, I decided to dig deeper.

That patient had a history of high blood pressure, and she'd had a stem-cell transplant that required her to receive chemo for preparation. Unfortunately, her myeloma came back after the transplant and she had received more treatments with chemo, daratumumab, and other medicines by the end of 2018.

The cancer then relapsed again.

After enrolling in our trial in October 2019, she received mAb85 on the first day. On the second day while she was in the hospital, her nurses noticed that minor bruise under the skin of her chest. They kept the patient in the hospital for observation but no treatment was necessary for it.

From this patient's blood work, it was apparent that her platelet count had decreased after the first day of treatment, coinciding with the minor bruise she had. The hospital reported the event as resolved spontaneously, and she was discharged the following day. The platelets were at a normal level by the time she returned for her next dose four days later. This patient continued to receive seven additional doses of the drug over the next few months, and at no point did her platelet counts veer from the normal range again.

That baffled me.

If the platelets were being destroyed by mAb85, I was confident the patients would have presented with clinical findings of active bleeding—it's hard to miss bleeding. Also the lifespan of platelets is about seven to ten days in humans. Yet the counts were getting back to normal within about seventy-two hours. If platelets were being killed, it was very unlikely that so many new platelets would be created by the bone marrow within such a short period. Again, I reckoned the platelets were not killed by the drug. Was it possible that the platelets simply moved away from the bloodstream to the tissue as a response to the new drug? It was difficult to prove such a thing. I consulted a platelet biology expert at the University of Pennsylvania. Upon reviewing the data, he was not sure what might be causing this unusual, short-lived thrombocytopenia.

Maybe the platelets were clumped together and miscounted? he offered. No was the answer. We had checked blood count charts from the hospitals. Were there high fractions of immature platelets? No. From the available data we could tell the platelets were not significantly immature. Perhaps there was an activation of certain proteins (complements) that caused it? Again, the data was not suggestive of that. We didn't see any evidence of complements activation, which can be measured in blood.

The biologist recommended extra laboratory tests to understand more but noted that it might be difficult to find an answer to this mystery.

The most plausible explanation I could come up with was this: platelet coverings (known as membranes) also express CD38, the target of our mAb. It was possible that the antibody attached itself to fractions of platelets floating in the bloodstream via the CD38 expressed on their coverings. Although this was not intended by design, this drug-platelet interaction via the CD38 might have caused the platelets to either move inside the tissue or be taken up by the cell cleanup system (known as RES).

This was far from a perfect explanation though.

Whatever the reason, we noted these issues in our safety observations and in our reports to the regulators. For my clinical colleagues who were enrolling patients in the trial, they were okay with this as it was an anomaly found only in laboratory counts and not clinically.

With additional tests and monitoring of platelet counts, we moved on with the trial. I was cautiously optimistic. But interruptions were on the way.

Chapter 10

A Symphony of Emotions

It would be possible to describe everything scientifically, but it would make no sense; it would be without meaning, as if you described a Beethoven symphony as a variation of wave pressure.

—Albert Einstein

FIRST APPROVAL

Spring 2020
Greater Boston

The party was in full swing at a fancy restaurant in Somerville by the time I reached there around seven o'clock in the evening. The music was on and folks were laughing and talking and having a great time. The aroma of seafood—New England clam chowder, lobster rolls, and fried delicacy—suffused the air as I stood at the corner.

We popped champagne for it was a night to celebrate success.

On March 2, 2020, the FDA had released their verdict on isatuximab: "Targeting cells has led to the development of important oncology treatments. While there is no cure for multiple myeloma, isatuximab is now another CD38–directed treatment option added to the list of FDA–approved treatments of patients with multiple myeloma who have progressive disease after previous therapies," said Richard Pazdur, director of the FDA's Oncology Center of Excellence and acting director of the Office of Oncologic Diseases in the FDA's Center for Drug Evaluation and Research. "In the clinical trial, there was a 40 percent reduction in the risk of disease progression or death with this therapy."

This approval was based on the data from the Phase 3 ICARIA trial—a much anticipated outcome for our group. The approval allowed doctors to prescribe isatuximab now together with pomalidomide and dexamethasone for multiple myeloma patients who had received prior treatments.

It was an invigorating first approval for our group.

I felt the energy within the room as I looked at the smiling and happy faces around me. Success can be hard to come by in drug making, contrary to popular belief. But here we were finally with an approval and gearing for the next stage.

Unknown to any of us, this would be the last time we would be gathered as a team for the next two years. Our lives were about to be changed forever.

A TRIAL IN SOLITUDE

February 2020
Lombardy, Italy

Thirty-eight-year-old Mattia started to cough with some shortness of breath. At the beginning it wasn't different from the other times when he had a flu-like illness, but his condition deteriorated rather quickly. He was admitted to a local emergency room in Codogno, a small town about thirty-five miles south of Milan. By the end of January, WHO had already declared the new coronavirus infection a global health crisis, the sixth global health crisis in the history of the organization.

Attending doctors at San Matteo Hospital tested Mattia for coronavirus. On February 20, the test results came back positive.

Was Mattia patient zero, the first documented case of COVID-19 in Italy? Or was he the second because a German visitor possibly had the infection on January 25? Regardless, Mattia had to be put on a ventilator as the oxygen levels in his blood plummeted.

While Mattia was fighting for his life, Mattia's dad died of coronavirus infection. Mattia's wife, very pregnant at the time, also got the virus. Thankfully she recovered. Mattia did not know he was spreading the virus to many before he fell sick—he couldn't have known.

"It's difficult to describe what happened to me. I remember hospitalization. I was in intensive care for eighteen days and then in the Infectious Disease Department, where I began to have contact with the real world and do the most beautiful thing: go back to breathe," Mattia said in a video message after he was discharged from the hospital.

Within days of his positive test result, Codogno and nearby towns in Lombardy were in lockdown, and on March 10, 60 million Italians were under a nationwide stay-at-home quarantine.

"We are out of time. We have to stay home," Giuseppe Conte, the Italian prime minister said. The region of Lombardy reported the highest number of COVID-19 infections and deaths in the days, weeks, and months that followed—a deadly trend that saw about forty thousand deaths due to COVID-19 in the region, the highest death toll in the country up to today.

Europe was in a state of shock.

Perhaps the Western Hemisphere was in a state of denial about the steps they needed to take to stop the viral transmission. There was confusion among leaders around the globe on the hard decisions needed to be made.

Talking to one of my physician colleagues who happened to be a proud Italian, I realized for the first time that the situation on the ground in Italy was desperate. They had run out of intensive care beds and ventilators. Doctors were asked to prioritize resources to those patients with the "greatest chances of survival."

Would you admit an elderly patient with coexisting health issues who might not survive long, or would you use your only available ventilator for a younger patient who had a better chance to defeat the virus? Such was the frantic, life-or-death situation inside Italy's hospitals.

It was almost unthinkable, yet this was the reality. And it was certain the pandemic would be deadly here in the United States. It was just a matter of time, but for some reason people in the United States weren't prepared for it.

On March 13, exactly eleven days after we celebrated isatuximab's approval in Sommerville, President Donald Trump declared a nationwide emergency due to the coronavirus and placed a travel-ban on travelers from twenty-six European countries.

There was confusion in our office that day. I was asked to go home and wait for instructions to return to the office physically. What about the ongoing trials? Should we pause them? No, the communication among the global centers was to be maintained via regular contacts. Officially, nothing was paused or stopped.

Eerie silence descended over the Charles as I drove back along its shore on Memorial Drive.

CANCER VERSUS COVID

Cancer doesn't care about COVID. It won't slow down because there is a pandemic going on and the world is in lockdown. It doesn't matter that all resources have been diverted to fighting the pandemic and healthcare professionals are occupied with managing a new, deadly virus. Cancer cells keep on spreading, invading new territory—tissues, organs, bones, and muscles—because that's what they are programmed to do.

This was a double-edged sword for our patients. They couldn't go for regular follow-up visits with their oncologists or even to receive their doses of cancer medicines because of the coronavirus measures. Cancer patients are often immunocompromised, and this meant they had a higher risk of becoming seriously sick if they got COVID-19. We had acute shortages of face masks and protective equipment. Doctors and nurses were working day and night. Due to the strict lockdown in certain regions, patients didn't have the chance to meet their sons or daughters or caregivers or loved ones who could accompany them to the hospital.

Then there was the dreaded reality of children and younger people spreading the coronavirus to their parents or grandparents. Social isolation and loneliness became the norm, most cruelly perhaps for our elderly.

Everyone stopped seeing each other in person.

Zoom became our best friend. For work we moved entirely to electronic video meetings with people connecting from wherever they were.

We kept clinical trials going although it became clear that many of the hospitals participating in our trials across the world wouldn't be able to enroll patients. I was receiving emails and calls from our local coordinators from different cities and countries saying that due to the overwhelming COVID-19 spread, they were in lockdown. Typically local staff would visit clinical sites regularly to monitor trial progress, review data, or answer any questions from the sites, but with the unprecedented surge of infections, this was no longer possible.

We instructed local staff to stop visiting the hospitals, following regional recommendations. Clinical protocols were modified where appropriate to ease the requirement of patients having to stay extended hours in the hospital for safety observations.

Enrollment in clinical trials came to a standstill. The Phase 1 trials were affected the most. Major hospitals and clinical research institutes only allowed enrollment in advanced Phase 3 trials and paused taking patients for new Phase 1 protocols. I kept hearing from experts across the globe about dire workforce shortages, as the doctors, nurses, and clinical staff battled the deadly infection.

In a video call to discuss our ongoing projects, I asked my colleague and myeloma expert Paul Richardson of Dana Farber Cancer Institute if he saw any signs of the pandemic slowing down.

"Not yet, my friend," Richardson replied. "If anything, I would like to commission you guys back to the clinical workforce!" He, along with several others, voiced the same concern: before it gets better, it is going to get much worse.

Right on cue, a dire situation started to unfold in New York City.

"IT'S APOCALYPTIC"

Facing an unprecedented wave of COVID-19, New York City was shut down in mid-March 2020. Hospitals in the city had never seen so many cases of a disease at the same time, and just like in Italy, and China, the healthcare system got swamped in a matter of a few days.

In Queens, one of the most diverse communities in the world where a fantastic mingling of culture, ethnicity, and social interactions make it a true global borough, Elmhurst Hospital Center became U.S. coronavirus ground zero. The hospital went over capacity—230 percent above its expected patient load with nearly all of these patients seriously ill with COVID-19 and gasping for breath.

Devastation. Chaos. And the smell of death hung over New York.

In a span of twenty-four hours, fourteen people died at Elmhurst Center. The morgue was full; staff waited for trailers to come and store the bodies. The medical staff had only a few dozen ventilators yet several times within a shift the loudspeakers would boom desperate requests for more ventilators as another patient was about to die in front of helpless doctors and nurses. Some died while waiting for a hospital bed.

The federal government sent a one-thousand-bed hospital ship to New York Harbor. Midtown Manhattan saw its convention center converted to a 250-bed hospital. A field hospital with white tents was set up in Central Park. The production of ventilators and protective equipment took top priority as local and federal governments scrambled to counter infection rates and deaths.

Throughout these moments of siege, I kept the communication going with our vast global network of cancer centers, hospitals, investigators, and country hubs. I held a weekly video call with blood cancer experts, their nursing staff, and coordinators to sometimes just listen to their experiences at the frontline. Week after week physicians reported no new patients from their sites for my trials. Some stated they were prioritizing large, randomized Phase 3 trials, which meant our mAb85 Phase 1 would not see new patients until the situation improved.

Although there weren't new patients enrolled in my trials, it felt right to still talk about how we might come out of this crisis. Several oncologists lamented that they had patients with advanced cancers who had missed critical appointments. They feared the interruption of treatment could mean progression of cancer for many. Alarmingly the rate of cancer diagnosis dropped by about 50 percent compared to the prior years, indicating patients were either postponing or canceling their appointments. Thousands of more cancer deaths due to postponement of treatment were predicted.

As the year went on, the number of COVID-19 cases started to decline. The death toll also tailed off. Virus testing became more available and new clinical protocols were put in place. The hope for vaccines was boosted by multiple private-government partnerships: finally, light at the end of a sinuous, dark tunnel.

On December 14, 2020, after a harrowing year of crisis, death, and despair, the first dose of a COVID-19 vaccine from Pfizer/BioNTech was administered to a New York–based critical care nurse. Four days later, a second vaccine, this one from Moderna, was also approved by the FDA through the emergency-use pathway. Some thought this a clear example of how very effective remedies could be brought to the public in record time when this was seen as truly necessary. The suggestion here was that for cancer medicines, the same perceived necessity to accelerate drug discovery was missing.

I don't agree with this sentiment. The COVID vaccines came out of earlier research in mRNA. Due to the pandemic, the regulators allowed super acceleration of these vaccine trials and their ultimate approval, on many occasions willing to allow greater risks to be taken. Anti-cancer medicines, on the other hand, require multiple disciplines to make a breakthrough. It is difficult to equate a novel infectious disease like COVID with cancers that have diverse reasons to form and far more than a virus. Of course, adopting more-efficient clinical trial designs with openness from regulators to embrace innovation would help accelerate cancer drug development.

Personalized vaccines against cancer using the same mRNA platform have gained momentum since the pandemic but there is still a lot to learn, especially on how a personalized cancer vaccine might be combined with standard treatment approaches.

The world celebrated the year end in 2020 in a subdued fashion. Gone was the spectacular fireworks display at iconic Sydney Harbor Bridge in Australia. Japan welcomed the new year quietly at home. Nightly curfews were imposed in France, Italy, Turkey, and Greece. In the confetti thrown at the Times Square's famous ball drop at midnight on December 31, 2020, were the wishes *love, connection, freedom.*

Nearly 2 million people had died globally from the novel coronavirus before the clock ticked to past midnight. This tragic burden offered me a perspective on cancer deaths, which are about five times more than peak COVID mortality every year, estimated to be about 10 million. I am relieved that we have very effective vaccines against COVID, but the quest to find a similar remedy for cancers is still a work in progress.

Chapter 11

"Not a Zero-Sum Game"

We have not wings, we cannot soar;
But we have feet to scale and climb
By slow degrees, by more and more,
The cloudy summits of our time.

—Henry Wadsworth Longfellow

CANCER CONSPIRACIES

There is a prevailing thought that the battle against cancer is largely ineffective. The success of the COVID vaccines prompts two questions in our minds: Could cancer drugs be created in record time? Is it that somehow there is a lack of interest in curing cancer?

This general notion ignores the fact that cancer isn't one disease. It is a collection of complex diseases that are perhaps as diverse as we humans are. I am confident that many of us know when it comes to survival rates from aggressive cancers there isn't a universal answer to the question of how long a person will live with a specific cancer. It largely depends on the cancer subtype, the time of diagnosis (the earlier the better), and the patient's response to treatment. Survival also depends on other variables such as age, comorbidities, and general health status. Then there are genetic factors, which are often hard to quantify in the role they're playing.

Can we do better in screening for and prevention of cancer? Absolutely yes. Can we improve our diagnostic techniques to detect cancers earlier? Yes, but with the same caveat that all cancers are not equal. While some cancers—especially cancers of blood cells—are relatively easier to identify using tiny markers of disease, this isn't true for solid tumors.

Are we creating better drugs for cancer treatment? This question is often passionately debated. I found myself as a target of this question many times.

During my tenure in Houston, I was lecturing at the University of Texas Health Science Center campus to a group of graduate students. Among them were a few clinically trained postgraduates who had signed up for the coursework that I was helping to teach. In a lecture concerning newer treatments, I was trying to explain the genetic markers of cancer and the idea of precision medicine: the approach of tailoring specific treatment to individuals on the basis of their DNA. I heard a murmur in the room, which was a bit unsettling for me, to be honest.

"Could you elaborate it a bit more?" requested a student from the second row.

"Sure," I said.

Several years before I had argued in an article that a patient-centric approach to treating cancer could improve clinical outcomes. I was excited to revisit the topic with the students in connection with precision medicine. Substantial challenges to precision medicine remain in practice. When designing new therapies, researchers need more evidence, especially on how cancer cells escape. Models that can mimic tumor behavior and its ways of evading drug actions are crucial. We need data from multiple different platforms to make precision medicine a reality.

"Think of precision medicine this way: each one of us, although so very similar in our genome, has a subtly different genetic makeup. How a specific cancer will behave and respond to a treatment is different between individuals because, technically, cancer cells are our own cells living inside us with our idiosyncrasies represented in them." I continued, "One drug may work well for a person but not so well for another. So the goal is to choose the drug or therapy that is most likely to work for the individual based upon hints: genetic mutations, biochemical markers, and other factors for that particular individual."

One example of this kind of treatment matching patients who have small cell lung cancer and a genetic mutation called EGFR with a tyrosine kinases inhibitor (TKI) drug class. By detecting the mutation before treatment, oncologists know these patients will benefit from TKI drugs over others.

"But isn't it about prediction?" the student asked. "Predictions can be wrong." From her facial expression, I could see she had difficulty accepting the predictive matching of treatment. It's a common reaction when I'm discussing treatment algorithms and biomarker-driven matching of patients with a treatment.

Some view matching a treatment in precision medicine like the predictive advertisements shown on our computer or smartphone screens by marketplace retailers based on our browsing history or our patterns of seeking information on the Internet.

But it is true that if a patient's genetic risk factors, cancer subtype, and cancer cells' sensitivity to different drugs are known then it's possible to offer a therapy most likely to benefit them.

The problem is we seldom have that kind of evidence available for a patient.

You may recall I started off this book with the Human Genome Project, which spent billions of dollars to map the (almost) full human genome. Craig Venter and others had hoped for a bonanza of new drug targets based on the genetic basis of disease. As it turned out, the variations in clusters of genes were too many given that any gene variant might increase the risk of a certain cancer. And that increase in risk, as determined by this microscopic variation, might even so be negligible or outweighed by other more macro-level risk factors such as advancing age. While there are mutations that we know have a direct relationship to treatment outcomes such as the EGFR–mutant lung cancer that I noted above, there are many other somatic mutations whose roles aren't clear.

Then there's the *microbiome*, the 100 trillion bacteria living in our skin, GI tract, and elsewhere. They influence the way our body responds to tumors forming and spreading. The microbiome influences genes and the expression of proteins, which in turn influences the *metabolome*—all the small molecules, proteins, fats, sugars, and other products involved in generating energy. All these processes are not stepwise, and one does not come after the other necessarily. They are dynamic, interacting events. If we study these processes in patients, all the -omics fields (e.g., genomics, epigenomics, proteomics, metabolomics), we'll have a better chance of tailoring our treatment to each individual.

The wealth of data is great, and it's integrating, understanding, and drawing meaningful conclusions from this data that's the critical task. But when a patient presents with cancer, having data at hand quickly to decide on therapeutics from a multi-omics point of view is not always feasible. But it may very well be possible in the future.

I tried to explain the complexity of new therapeutic approaches to my students at the risk of going beyond the scope of the lecture. But they wouldn't let go, they wanted to know how we screen drug candidates.

"You can study the effects of several drugs on the cancer cells first, in the laboratory, and know what will work and what won't, no?"

"We can, and we do sometimes," I smiled at the questioner. "But it is more complex than that. You can obtain cancer cells from the patients and test the effects of hundreds of molecules in the laboratory, but there is no guarantee that these results will hold. Cancer cells outside the human body or a biological system aren't the same in their behavior as in the laboratory plates."

"Then find a better disease model!" quipped another student, a clinician. I was relishing this discussion as I felt these were important questions for drug discovery and development.

"Agreed, we need to find models that are better at representing cancers," I said. "But remember, models are almost always wrong. You can't find an animal model or other type of model that is 100 percent predictive."

"You can create better drugs," said the girl who spoke first. "You can have treatments that can cure cancer. But wiping out cancer isn't profitable— rather, keep the market with one new drug here, one there!"

I noticed the bitterness in her tone.

Maybe she'd had personal experiences of dealing with cancer. Perhaps a close family member, someone she loved, had had a difficult journey. I often hear this sentiment in my line of work: that the corporate world could cure cancer if they wanted to but it would stop the flow of money for them.

This is not true.

Assuming all cancer could be cured ignores the complexity of cancer and its diversity. A universal cure of cancer is likely not to be found. Perhaps a better goal would be to focus on curing one type of cancer, such as a B-cell lymphoma. Or striving for longer survival from acute leukemia. Or perhaps something different: improving the quality of life of our patients living with multiple myeloma. Yes, I chose certain blood cancers that I'm familiar with (conveniently), but the point is, by choosing to eliminate specific cancers or by improving survival in certain cancers and by helping patients achieve better quality of life with cancer, we can reduce the lofty cure-cancer goal to several achievable goals.

I don't believe a scientific breakthrough of magic bullets for cancer has been discovered but shielded from daylight. It doesn't make sense or seem feasible to hide such a huge achievement. For one thing, cancer research is a collaborative process. Hundreds of people are involved in the research and production of a drug. It's difficult to see how any company could keep such a secret. But if profitability is supposed to be the motivation, a cancer cure would make a colossal amount of money.

HOW CANCER ESCAPES

It is an unfortunate reality that with time, cancer cells acquire the ability to withstand drugs, which eventually lead to relapses. In most patients with a cancer that is R/R, we need alternative treatment plans. The backup plans largely depend on our understanding of how cancers become resistant to drugs.

Tumor cells form and live in a biological scene with other actors. The cells need to feed, and they get nutrients from the tiny blood vessels that branch off from a larger supply chain of blood vessels. The tumors also have immune cells—B-cells, T-cells, NK cells, dendritic cells—in their vicinity that affect their growth. In this scene, the tumor cells rest on an organized, stable mesh that supports them called the *matrix*. Some matrix-forming materials create a dense layer, a physical boundary called *basement membrane* that surrounds specific tissues, providing support, balancing the entry of supply chains (i.e., blood vessels), and preventing bad actors like tumor cells from invading or spilling over and invading surrounding tissue.

These are cancer cells' own milieu. These microscopic players influence the early growth and development of cancer cells.

There are other, greater drivers of cancer dictated by the genome. And of course, when we treat patients, the pressure of the drugs on cancer cells lends a major part on their evolution. All of these together play out in a dynamic battle that ultimately determines if a cancer will become resistant to treatment. Today, much emphasis is given on the roles of different players in a tumor's microscopic environment to understand how cancers form, and how they evade drugs. It's fair to say, we still need to learn a lot more about these minute cancer boroughs which would help scientists to put an end to therapeutic resistance.

In the case of multiple myeloma and CD38–targeting antibodies like daratumumab, the resistance mechanism is intriguing.

Since CD38 is the target antigen, once patients receive daratumumab or isatuximab, myeloma cells that wear the boldest of CD38 kits (overexpressed CD38 on their surface) get killed first. This killing happens quickly.

As soon as the first dose of daratumumab is given, a nearly 90 percent drop in CD38 expression is observed in myeloma cells. Along the same line, when a patient progresses from treatment with daratumumab, CD38 is found to be low on their myeloma cells. This low CD38 situation persists for about six months after the last dose of daratumumab has been given. If the target is no longer present or lowly expressed, where will our weapons hit? A low CD38 expression status can also be found in a patient with a great response to the drug.

On the other hand, CD38 antigen is expressed by some normal immune cells too. Some immune cells are destroyed by our drugs, an unintended but unavoidable event giving an advantage to myeloma. The cell-eaters in our body—the phagocytes—that are supposed to capture cancer cells also get fooled by don't-eat-me kits worn by the myeloma cells as a response to the attack by anti–CD38 antibodies. All these reactions, adjustments, and refinements contribute to drug resistance in multiple myeloma.

When it comes to other cancers, they are equally as creative and flexible in escaping the actions of drugs intended to obliterate them. These cells deploy multiple strategies to survive—it's a life-or-death situation for them too, after all.

HASTA LA VISTA

Spring 2021
Greater Boston

I received a call from the company vice president in the late afternoon. We were in the office now three days per week, and on that day I was working remotely.

"I just landed here in Boston, let's catch up tomorrow" was all he said. I sensed it was something urgent. About a week earlier we had monitored the progress of the mAb855 trial. Notwithstanding the pandemic, enrollment was okay. I told him that I'd be in the office the next day.

We met in our brand-new building in Cambridge the next morning.

"We need to review all data of the mAb85 trial with the CMO," he said.

"Not a problem."

I was ready as I had just gone over the data with a group of myeloma specialists: global thought leaders on the disease. There was not a lot to share as this was still a Phase 1 study with possible RP2D all but reached.

Later in the week, the CMO and senior leaders of the company joined via Zoom as I and the vice president went over the data. I spoke briefly about enrollment: "Considering the pandemic, we are fine with the enrollment. There is a lot of interest in the network now that we are almost at the end of dose escalation."

"Can you compare efficacy between mAb85 and the first-generation CD38 antibodies?"

This was predictable: mAb85 targeted CD38, the same antigen as isatuximab and daratumumab.

But there were nuances.

"While I can try to compare," I said, "it won't be an apple-to-apples comparison."

The difficulty for me was that the two drugs treated patients at different stages in their disease. The first generation of CD38 monoclonal antibodies largely enrolled patients in the trials who had not been treated with CD38 drugs before—we called them *CD38 naive* patients. On the other hand, almost all in the mAb85 trial were R/R myeloma patients: they'd received prior treatment with either daratumumab or isatuximab in addition to other treatments.

We've seen the drug resistance patterns of myeloma cells: they reduce the target (CD38) when treated with daratumumab. At this stage when the myeloma have low targets on them, it's harder for another drug of the same class to put up a good fight. To make things even more complicated, the doses were different between the first and second generation of CD38–targeting antibodies. The mAb85 doses were relatively lower than either daratumumab or isatuximab. The way the drugs were infused also differed between them.

These findings from individual trials of the second- and the first-generation antibodies were difficult to compare. As the meeting progressed, it turned out we needed more specific comparisons between the drugs. In my mind, it was not just an apple-to-apples comparison, but now we needed comparisons between different varieties of apples!

More data was needed for a clear understanding on whether mAb85 was better or how much better. It was certainly not worse than the first generation in early data, but not being worse rarely cuts it in drug development. About a week after our review, I was notified of the decision to stop developing the program due to strategic considerations.

I was disappointed and despite the heads up couldn't help feeling low.

It was early spring in Boston though winter still gripped us firmly, and springtime depression sure felt more severe that year than in other years.

My usual drive home along the Charles failed to improve the mood.

"TUMOR-METER"

I like to compare diabetes and cancer a lot, much to the irritation of some of my teachers and mentors over the years. In my defense, both are chronic illnesses with a trend of increasing new cases globally over the past decades, and both are expected to be among the leading diagnoses in the world with an estimated one-half billion people living with either diabetes or cancer by 2040.

The similarities don't end there. Cancer and diabetes share risk factors: smoking, obesity, poor diet, and physical inactivity, to name a few. The risk of getting cancer increases with a diagnosis of diabetes—with some cancers the risk almost doubles. Diabetes damages the immune cells, B- and T-lymphocytes, that are key players in several lymphomas. Thus the risk of T-cell lymphoma and leukemia increases with having diabetes. This much we know.

Despite many similarities, one thing that distinguishes the two drives me crazy.

In a job interview a few years ago I was going off on a tirade about cancer and diabetes when the professor stopped my rant and asked, "What's your point? That we oncologists don't do a good job of detection and screening?"

"No, they [diabetologists] have something we don't: a glucometer!" I replied. "They can measure blood glucose instantly and plan treatment accordingly. I dream of the day when we'll have a matchbox-size device that can test my blood and tell me in real time if I have cancer in my blood!"

He laughed. "A single drop of blood to run hundreds of tests in the office and diagnose cancer—sounds very familiar!"

The professor was alluding to the bold claims made by the Silicon Valley biotech Theranos and its charismatic, nineteen-year-old founder Elizabeth Holmes. She claimed that a portable device Theranos made could run dozens of diagnostic medical tests on a single drop of blood and detect disease. It could not. The disgraced company, which at one point was valued at $10 billion, closed its labs in 2018. Holmes, along with the former president of Theranos, Ramesh Balwani, were both indicted for fraud and later convicted by juries.

I joined in the laughter with the professor. Indeed, my outlandish vision of real-time cancer detection in blood—a tumor-meter, sounded very similar to what Theranos had promised.

Except this is becoming a reality in cancer now through a technique called liquid biopsy. No, it's not from a single drop of blood nor does it have a matchbox-size meter to display the results at the bedside. It's done through laboratory tests and the results are interpreted and reported by a medical team with expertise in oncology, pathology, and genomics.

Nonetheless, this method is beginning to allow oncologists to search for remnants of cancer cells in the blood when conventional imaging—X-rays, CT scans—can't detect it. This method of tracking cancer is known as *minimal residual disease* (MRD) detection, and it's taking the cancer world by storm.

To me, MRD tracking by liquid biopsy represents a tremendous tool for overcoming drug resistance in cancer. We all agree that the best way to address drug resistance is to detect cancer early when the burden of tumor is low and when the cancer cells haven't quite deployed their evasion strategies through mutations. MRD detection methods can identify precancerous conditions, which can be treated before they become full-blown cancers. It can also tell a patient and his or her medical team after a course of therapy how deep the patient's response is to the therapy.

Conventional ways of cancer monitoring are based on serial radiographic imaging to measure the change in size of a tumor. While it can detect progression of disease, imaging is not useful in monitoring how the tumor is changing in response to therapy. But real-time tracking of MRD can offer a heads-up that a patient may relapse soon. That in turn gives us an opportunity to modify or change treatment approaches.

There are several ways to detect MRD, but one of these captivated me most due to its sheer ingenuity: it is called *circulating tumor DNA* (ctDNA) tracking. Once again, genomics is at play here.

It started with a New York–based scientist, S. A. Leon and his colleagues' assumptions that in cancer patients' blood, higher levels of DNA might be expected given there are growing tumor cells in their system. Cells suffer from injury all the time both in normal conditions and at times of disease. DNA resides inside cells, and when a cell is destroyed, pieces of its DNA are released into the bloodstream. These DNA remnants are cell-free DNA floating in the plasma.

While cell-free DNA could also be from non-cancer cells, patients with any one of several cancer types—lung, breast, colon, and lymphoma, among others—have greater amounts of cell-free DNA in their blood than expected.

Leon and colleagues suggested in 1977 that if cell-free DNA persists at a higher level after cancer treatment, this can mean the patient had a poor response to it. On the other hand, if cell-free DNA levels decrease after treatment, it likely means a good response to the treatment.

"We hope, therefore, that sequential measurements of DNA concentrations may be a useful tool for monitoring the effects of the therapy," they concluded in their paper.

But at the time they didn't have the technology to distinguish between tumor DNA and normal cell-free DNA. Serial measurements of cell-free DNA wasn't feasible.

The challenge was to determine whether the cell-free DNA had come from tumor cells or normal cells. That's where newer developments in human genome sequencing have proved to be very useful. By comparing a patient's normal tissue with their tumor biopsies, we can compare the two DNA sequences.

Once that is done, the cell-free DNA bathing in the patient's blood is magnified through an established method like polymerase chain reaction (PCR), and researchers can then determine if any ominous calling cards of the patient's former tumor are present. All of this happens at the molecular level. This can be measured on multiple occasions since it only requires a blood sample from the patient, and the burden of cancer can be monitored over time this way.

There are limitations and challenges to circulating tumor DNA tracking, mostly because of the technically demanding way the blood samples need to be collected, prepared, and analyzed. But there is no denying that real-time monitoring of drug response will allow oncologists to modify dose or schedule or even pick an alternative treatment plan for their patient. By monitoring ctDNA we can pick up mutations in cancer cells earlier, and this can save valuable time when patients need to receive an alternate treatment before it's too late.

I am convinced I see my tumor-meter already. Well, yes, it's not quite a meter per se, but I think MRD detection is as close to it as possible.

Like the father of vaccines, Edward Jenner, who predicted what was in store for smallpox after his vaccine discovery, I would like to do the same by saying the annihilation of drug resistance in cancer, the most dreadful outcome of a therapy, must be the final result of MRD monitoring!

But it's not a zero-sum game after all. Every new drug, therapy, or way to fight cancer counts. Looking back, I can't quite say I disagree with the decision to pull mAb85. It is always a challenge to pick and choose your assets, and which one to push forward or which one to close sometimes depends on a broader strategy.

And I had other battles to fight, new treatments to be created and developed.

In this fight, we now have a shiny new weapon called *cell therapy*. From pioneering scientists whose lives were uprooted by a major war to a Navy scientist who left research in HIV to pursue the fascinating world of T-cells to five-year old Emily whose aggressive leukemia was confronted by this novel breakthrough, and from the perils of unleashing our own immune system to the magical phrase *You are cured!*—a living drug may change it all.

PART III

Riding the CAR

Chapter 12

The Rise of T

The opposite of war isn't peace, it's creation.

—Jonathan Larson

FROM THE ASHES

Spring 1944
Budapest, Hungary

It always fascinates me how destruction and violence connect with creation. The devastation in times of war seems to spark creative energy. The evidence of this was everywhere during and in the immediate aftermath of World War II. Our knowledge of the mysterious ways of working of the human immune system may have been shaped by this conflict as well.

It was early hours on March 19 in Budapest when Hitler's army invaded the city. The invasion was called Operation Margarethe and was to occupy and take control of the Hungarian territory. German forces wasted no time in marching from Austria per the order of the Fuhrer.

At the start of the war in 1940, Hungary had joined Italy and Japan to form an alliance with Germany—the Tripartite Pact, as it was known. By June 1941 they had declared war against the Soviet Union and sent thousands of soldiers to the Eastern Front.

The Red Army crushed them on the frontline.

At the battles of Stalingrad and Voronezh, Hungarian forces were effectively wiped out by the Soviets. The belief in a German victory started to fade away, and the Hungarians wanted to make peace with the Allies. But without its armed forces, the country lay at the mercy of Hitler.

The Soviet Army was closing in on Budapest.

The Fuhrer had a sinister plan. He wanted to punish some eight hundred thousand Jews living in the region. Until then the Jews in Hungary had been largely protected from the detention camps by virtue of the Tripartite Pact. Not anymore. Hitler decided it was time to force Hungarian Jews to submit to the German authorities.

Fear and terror engulfed the banks of the Danube as mass deportation of Jews to Auschwitz began in Budapest. Pal Teleki, Hungary's prime minister, was deeply ashamed of his role in the alliance with the Third Reich. "We've thrown away dignity. We have allied ourselves with the scoundrels and I did not hold you back—I am guilty," he wrote in a note to his countrymen and then committed suicide as the Germans entered the capital.

Nineteen-year-old Jewish-Hungarian George Klein was caught in this turmoil. At the time he was serving as an assistant secretary at the Jewish Council of Budapest. Klein's family originally moved from Eastern Slovakia in 1930 to flee discrimination and yet fourteen years later in Nazi-occupied Hungary, their fate was all but sealed.

Because of his job at the Jewish Council, Klein had access to a secret document written by Rudolf Vrba and Alfred Wetzler, who had managed to escape the concentration camp at Auschwitz. Klein's manager at work allowed him to read the descriptions these two men had written what was known as the "Auschwitz Report" and share with his friends and family what was happening in the camp.

Most Jews in Hungary did not believe the report. How could this be true? Murdered in gas chambers in the name of resettling? But young Klein believed it must be the case. Why would Vrba and Wetzler lie? They only wanted to warn Hungarian Jews of what lay ahead for them once they'd been deported by the Germans.

Klein was arrested by the Nazis soon after.

They put him into forced labor as he waited for his train to Auschwitz. He kept thinking about the report. Klein knew boarding that train meant certain death.

We don't know how exactly Klein escaped, but when he was ordered to board the train, he managed to run away. Klein hid in a cellar until January of 1945 as the terror against the Jews continued. About half a million Hungarian Jews were deported to the Auschwitz concentration camp from Budapest, most to their deaths.

On January 10, 1945, as the sun rose over the Danube River, Klein emerged from a cellar on the outskirts of Budapest. Before him lay the ruins of the city he had known as his home: shattered houses and the corpses of soldiers, civilians, and horses all tangled up in death.

Klein had mixed feelings.

"I suddenly realized, with surprise, guilt, and delight, that I had survived in spite of an 80 percent chance that I would end my nineteen years in the gas chambers or in a military slave labor camp," Klein recalled years later.

Budapest was still under siege.

The Russians took command of the eastern side of the city. They patrolled the area alert for disguised Germans. Desperation and hunger drove people to the streets where many were caught by the Russian army. For Klein, fleeing the Russians was nothing compared to his earlier escape from a Nazi labor camp.

Most of us wouldn't think about formal learning or education at a time of devastation like this, but Klein was different. In the liberated part of Budapest, he strolled around and quickly decided he would attend medical school. Perhaps the death and suffering around him inspired Klein to become a healer, it's hard to say.

Once the streets were open for the public again, Klein walked straight to the University of Budapest only to find deserted buildings and the bodies of soldiers in the rubble. But Klein and a friend were adamant: they must go to the University of Szeged, where they would get a chance to be admitted to medical school.

Located on the southern plains near the Serbian border, today it takes only about two hours to reach Szeged from Pest. For Klein and his friend, it took more than five days to reach the University of Szeged. They walked, hitch-hiked on horse-drawn carriages, or traveled by whatever means were available in the war-torn country, even on a Russian military truck. Their persistence paid off: the duo reached the university on a cold February morning and were admitted to medical school on the same day.

Klein noted the strangeness of the environment at Szeged.

There weren't any professors: they'd fled the country during the past year or so. One assistant professor taught everything—anatomy, physiology, pathology, and forensic medicine, which had been his area of expertise originally. The students who enrolled were also war-scarred. Some had come out of hiding places like Klein, some from detention centers or labor camps.

Typically, first-year medical students start off with dissections in anatomy class. Here there was no shortage of dead bodies. The dissection hall had piles of cadavers drenched in pungent formalin that irritated the eyes and partly dissected bodies lay on the tables.

This was the perfect place for Klein.

Finally he had access to what he described as an "enchanting landscape, a previously forbidden paradise." Days and weeks passed by as Klein immersed himself in the study of human dissection and examination.

After three semesters at Szeged, Klein returned to Budapest to continue his studies in the university there. There were few resources, but he managed to collaborate with an internationally recognized professor of cell biology there, and later he went to work in the Pathology Department both as a student and an instructor.

In summer 1947 on a rare week-long vacation on Lake Balaton on the terrace of a bombed house, Klein fell in love for the first time.

LAKE BALATON

Professor Balo was much irritated by Klein's tenacity at work.

He was Klein's supervisor at the Department of Pathology at Budapest and knew about his student's workaholic nature. The laboratories needed a fresh coat of paint but Klein refused to leave the tiny nook he was working in, refusing to let anyone paint it.

"You can take vacation for two weeks for once!"

Professor Balo gave his marching orders. It was like falling from paradise for Klein. What was he going to do for two whole weeks? Frustrated, he tagged along with a few students planning to spend a week on Lake Balaton. They would stay in a lakeside deserted villa, the couples in the group planning a romantic getaway.

The group spent the first night on the terrace of the building. Klein found it pleasant, unexpectedly so, but he didn't know it was going to get a whole lot more interesting. On the second day, his fellows went to the train station to receive another student who was joining the gang on the trip.

Eva Fisher was a student at the Department of Pharmacology at the University of Budapest. As she was walking with the other boys up the hill from Lake Balaton, Klein saw her.

Eva radiated beauty in a somewhat strange way. Her eyes expressed both delight and ache, both gravity and folly as she greeted Klein. He now recalled where he'd seen her before: at the University in Szeged by the door of the dean's office, for the first time. And then he'd seen her name on posters as an actor in various plays. Eva had attended some of Klein's autopsy demonstrations as well, Klein vaguely remembered. He was too occupied to notice the girl then.

But this time it was different.

The couple fell in love in that ruined villa. The passion of it consumed workaholic Klein to the point that all other interests and problems appeared to have vanished as "if they never existed."

They spent eight days at the lake, enthralled by love.

Reality beckoned on the last day of the vacation. Klein received an invitation to visit Sweden, against all odds arranged by one of his students. At the time it was hard to get an exit permit out of Hungary from occupying Soviet forces.

This was a dream come true for Klein.

Except he did not want to go to Sweden anymore. He'd just met Eva, and leaving her now seemed impossible. Yet Klein boarded the train on a Sunday morning with a heavy heart. Eva couldn't help thinking she'd seen Klein for the last time.

From magic to melancholy, summer 1947 was one of the most happening periods for the Klein couple. They would be reunited two months later when Klein came back to leave Hungary permanently. A Communist takeover seemed destined, and young scientists were advised to get their degree as soon as they could and leave the country. There was nothing for Klein in Hungary other than the girl of his dreams, Eva.

Klein and Eva got married secretly in August 1947 in front of two friends. The registrar was the only other person present. The newly married couple had to keep the event confidential as they needed to wait for the right moment to leave Hungary.

It took six more months for them to finally share a place together as husband and wife and continue their studies to obtain medical degrees. Sweden became their permanent home.

THE ELUSIVE IMMUNITY

In chapter 11, we met the matrix of tissue and blood vessels where tumor cells coexist with immune cells. Once a tumor starts to form, the dynamic interplay between tumor growth, new blood vessels supplying the growth, and the behavior of the immune cells in the tumor milieu determine whether the tumor will progress on its journey to become a cancer or not. This phenomenon of our immune systems affecting cancer formation came to be known as *immune surveillance*—a hot topic in present-day cancer research.

Our knowledge of immune surveillance wasn't rich in the World War II era when Klein and Eva moved to Stockholm. The antibodies and their role in immune defense were understood after the work of Ehrlich, Kitasato, and von Behring, but there was more to the immune system than antibodies. At this time, Ciser Milstein was pursuing his doctoral studies in Buenos Aires and yet to explore immunology research. Several hypotheses about cancer causation and the role of our immune system in it had been offered, but the immune response was still an elusive topic.

Paul Ehrlich declared in 1907 that in a host, cancer cells are strongly challenged by the host's defensive acts, which in his words were "positive mechanisms of the host." And because of these positive processes, unusual or precancerous cells don't usually become cancers. In the same year, Ehrlich discovered a cancer originating in the breast of the white mouse strain. These cells were called Ehrlich's Cell. The peculiarity of these cells was that if they were injected inside the abdominal cavity of another healthy mouse, they created tumors and irritation resulting in a collection of a liquid inside the abdomen. The liquid contained free cancer cells bathing in it. This cancer-cell-containing fluid was so potent that even if it were diluted and treated with anti-cancer drugs, it still remained cancerous.

The fluid came to be known as Ehrlich ascites carcinoma (EAC) and is still used as a tumor model in cancer research. *Ascites* means fluid collection in the abdomen—a manifestation of liver damage, among other things. *Carcinoma* is a cancer of the cells that provide a covering for our organs.

As young student-scientists at the Karolinska Institute, Klein and Eva studied EAC and asked the question: Why is Ehrlich's ascites carcinoma unique? If these liquid tumors spread so easily, why don't other tumors do the same? Possibly most cancers needed a milieu of solid tissue to grow and not a free-fluid form, the young scientists posited.

To make use of the Ehrlich ascites carcinoma model, Klein and Eva needed experimental mice, which were impossible to obtain in Sweden at the time.

The impossibility became possible. Klein got an opportunity to work with and learn from American geneticist Jack Schultz in Philadelphia at the Fox Chase Cancer Center. During his four-month visit to the United States, Klein learned a great deal about genetics and working with mice. He tried growing solid tumors in a fluid in the abdomens of the mice. At the end, Klein returned to Stockholm with renewed vigor for learning and asking questions—and with two hundred live mice.

Armed with mice ready for experiments, Klein and Eva succeeded in creating fluid variants of solid tumors at the Karolinska Institute. These ascites tumors the scientist duo grew appeared to be more aggressive. They could be passed between multiple generations of mice. This subset of tumors was different from original cells owing to variations. This meant that tumors in this ascites form could spread more easily in a host possibly because the immune defense of the host mice had been overridden.

On the other hand, if the host was exposed to weakened (attenuated) tumor cells first, the host developed an ability to reject tumor grafts later—a vaccine concept had emerged from Klein and Eva's experiments.

How does a host become immunized from tumors arising from its own cells?

In 1960, Klein and colleagues showed that host mice can reject chemically induced *sarcoma* (a cancer of the soft tissue) if they're exposed to these tumor cells previously. But this apparent immunity of the host was specific to the original tumor only—it wouldn't reject another type of tumor. There was rarely any cross-immunity here unlike Edward Jenner's antibody experiments where we've seen cowpox exposure gives immunity from smallpox. This specific defense against a specific tumor by the host when previously exposed to it appeared to be orchestrated by players in the immune system other than B-cell-producing antibodies.

At the same time in the early 1960s, a young physician-researcher in the UK was working in a makeshift laboratory in a horse-stable in Kensington hoping to explain the role of a "leftover" organ in humans. He too had escaped that fateful war.

"A USELESS ORGAN"

August 1941
Shanghai, China

The young family of four was still grieving the loss of their teenage daughter Jacqueline. With the burden of grief, they had to leave the country secretly as the political situation in China favored the Nazis and their puppet regime in Vichy, France.

The Meuniers were French by birth.

Maurice Meunier, the father, spoke English, Spanish, Mandarin, and Japanese—he was an asset when it came to being an interpreter. When war broke out in spring 1939, the family moved to Shanghai from Lausanne anticipating the instability of the Swiss region from Nazi aggression. Meunier was the manager of a French-Chinese bank in Shanghai at the time. But he was a supporter of Charles de Gaulle, an opponent of the regime in France. Due to this and the imminent threat of neighboring Japan joining the war, he needed to leave China to avoid being arrested.

Meunier cut a deal with the British.

The family got new passports with a new surname: Miller. They boarded the last cargo boat out of Shanghai harbor for Jakarta, sailing from there on a passenger ship to Sydney.

Jacques Miller, the youngest of the siblings, hated the war. Like Klein he was motivated to serve the wounded in that time of violence. The fact that his elder sister Jacqueline had contracted tuberculosis but he and his younger sister Jeanine did not despite their living in the same house intrigued Miller. He was committed to study medicine and perhaps conduct research on immunology if he had the chance.

Miller completed medical school in Sydney and applied for a research fellowship in the UK. He arrived at Chester Beatty Research Institute in

South Kensington in 1958, the same year Ciser Milstein joined the Medical Research Council to embark on his antibody research.

Miller was interested in studying cancers such as lymphocytic leukemia that are induced by viral infections. Chemically induced cancers did not appeal to him much. Miller wanted to explain mechanisms of cancer formation in a biological model. In lymphocytic leukemia, he got the perfect project.

In these virus-influenced tumors, the cancer cells develop in an organ known as the thymus and then spread to other parts of the body. Now this organ, the thymus, was a gland (meaning it secretes substances) thought to be a leftover structure in our evolutionary journey with no useful function. Think of it like the wings of birds that can't fly. When I was in medical school, the thymus would often be compared to the appendix, an organ with poorly understood functions. The thymus sits behind the sternum in between the lungs.

If mice are injected with a certain virus right after their birth, a particular leukemia starts to form, but later introduction of the virus doesn't produce a tumor. Miller thought the virus possibly multiplied only in certain cells in the thymus of the newborn mice. Once the thymus was more mature in an older mouse, the cancers wouldn't form.

Miller designed a surgical experiment.

In a converted laboratory in a stable in Kensington, he began with three groups of mice. One group received no special treatment while each mouse in the other two groups received injections of the virus within sixteen hours of birth. After about a month, Miller removed the thymus surgically from half of the mice that had received the virus while the other half were left with their thymus intact. The animals were observed side by side in cages.

The results were unexpected.

The mice that received virus injections but had their thymus intact died of leukemia the fastest: within three to six months. The mice that received no special treatment (no virus injections and intact thymus) also developed leukemia, but later in life at the age of seven months or more. Curiously, the group of mice that Miller had injected with the virus at birth and then removed their thymus soon afterward did not get leukemia.

The conclusion from this neat experiment, which Miller published in *Nature* in 1959 as the solo author, was that removal of the thymus in mice in early life stopped them from having leukemia later in life although they were exposed to the virus at birth. It indicated that the leukemia cells needed the immature thymus cells to grow.

His next move was to transplant (or graft) thymuses obtained from newborn mice to adult mice after removing the adults' own thymuses. Inevitably all the adult mice developed leukemia after they received a newborn's thymus graft. The viruses injected in the mice remained dormant in absence of a thymus but were activated after the thymus transplants—a cool experiment that needed sophisticated skills in surgery.

Miller went on to repeat his studies in the National Institutes of Health Science laboratories in Bethesda in 1963 in germfree tanks instead of a converted horse stable. Additionally, he grafted foreign tissue (skin) taken from other mice and rats onto mice without thymuses and compared them to normal mice with intact thymuses. The normal mice rejected the foreign skin graft—an expected immune response—while the thymus-deficient mice failed to do so. This was the clearest evidence yet that in order to develop an effective immune response against "foreign" tissue, a thymus was required.

Once thought of as a useless, leftover organ or a graveyard of dying cells, the thymus proved to be critical for the development of immune cells in early life.

Later, Miller and his student Graham suggested that there were two different cell types that participated in the immune response. One cell type produced antibodies (later known as B-type lymphocytes) while the other, smaller type was called a T-cell (to indicate its thymus derivation). T-cells cooperate with B-cells in the body's immune response.

Some scientists of the time didn't like this explanation at all. Among them was a professor of immunology at Curtin School of Medical Research in Canberra. In his opinion, B- and T-cells represented only the first and the last letters of the word *bullshit*!

He was wrong. Soon immunologists across the world had jumped on the T bandwagon.

The rise of T had begun.

Chapter 13

A Killer Within

The value of an idea lies in the using of it.

—Thomas A. Edison

(REVERSE) NAMESAKE

Spring 2023
Greater Boston

It was accidental that I sat next to the gentleman at a scientific symposium in Boston. I arrived late, the room was already full, and someone pointed me to what appeared to be the only empty chair on the left side.

So I took it.

During the course of the next two hours or so, the organizer of the symposium covered newer developments in anti-cancer drugs. The science was sometimes difficult to follow as we jumped around from small molecules to antibodies to multi-target antibodies.

I remember asking the gentleman sitting next to me about his thoughts on the multi-target drugs that were being presented on the screen.

"Whatever," he said with a wicked grin. "I am a TCR guy!" *TCR* means *T-cell receptor.*

Next was my turn to speak briefly to the audience, so I missed the implication of what he'd said, but when I returned to my seat, I glanced in his direction. Seen in profile, his face certainly resembled that of a traditional samurai warrior with fire in the eyes and a goatee.

"Hi, I am Mak," I introduced myself.

"I am Mak, too!" He said with the same amused grin. Turns out, my first name was his last.

He was Tak Mak—the discoverer of the structure of TCRs, which he did decades ago at the age of thirty. The field of immunology, especially cell treatment, has greatly benefited from Mak's pioneering works.

What followed was an interesting discussion throughout the day that led to a visit to a diner nearby. Over a humongous Tomahawk steak and a shrimp tower, I listened as my namesake talked about his brilliant work, injecting his frank wit on science and his views about the world as we know it.

In the beginning Mak was cleaning test tubes in a laboratory for $1 an hour. When he was offered a raise to $1.50 an hour if he was willing to conduct research, Mak jumped on it. This was Roland Rueckert's laboratory in Wisconsin circa 1965.

The concept of cell receptors was first offered by Paul Ehrlich in 1897. He noted that all active substances function by attaching to a structure on the surface of the cells that was, in his words, "side chains." Once a substance like a toxin attaches to these side chains, it acts upon it and the chain structures are released. The cells generate new side chains to take the place of the old ones, and during times of stress the cells produce more of these chains, which are then shed into the blood to protect the cells from future infection. Ehrlich later dubbed the side chains *receptors*. He described detached receptors as antibodies.

While the side-chain theory had its shortcomings, the concept of receptors was groundbreaking. These are chemical structures that are proteins, and they can receive and convert signals in a biological environment. The receptors present on the surface of B-cells were heavily featured in Ehrlich's side-chain concept. But there are other types of receptors that are located inside a cell.

When Mak started to conduct his own research at the Ontario Cancer Institute under the supervision of Ernest McCulloch in 1972, his idea was that because B-cells and T-cells function so differently—one type (B-cells) produces antibodies, the other (T-cells) fight invaders—their surface receptors must be different. For him, TCRs were the "main antenna of the immune system. They can detect viruses, cancer cells, without affecting the normal organs." So the TCR does the detection, recognition, and presentation of foreign threats.

Mak, his postdoctoral fellow Yusuke Yanagi, and a technician set out on a mammoth task: identify T-cell markers and hopefully clone the elusive genes of the TCR. In their first set of DNA sequences from the experiments, they succeeded in identifying one of the TCR chains, called *beta chains*. Independently, Stanford's Mark Davis group also achieved the same results, and the two teams published their findings in 1984.

Within a year the complete structure of the TCR had been revealed. It is a protein composed of two chains linked by a disulphide bridge: the same bond that Ciser Milstein thought offered diversity in antibodies.

Mak likes to call TCRs the key to a car. In this analogy, a protein expressed on the T-cells, called CD28, is the accelerator, driving the cell onward, while another smaller protein called IL2 (and IFN) supplies the gas. The brakes that slow down the vehicle in this example are yet another protein called CTLA-4. Together, these dynamic interactions influence the ability of T-cells to mount an effective immune response via TCRs.

With his colorful skills in oration, Mak says the T-cells are superheroes. Like James Bond, they can find and eliminate the enemy. We only need to help these spies in blood target their nemesis.

SELF RECOGNITION

The way TCRs operate is complex. In the thymus, immature T-cells are checked to make sure they don't turn against the self. Here *self* refers to the native host, the human being. If TCRs have too much attraction for self proteins, they are programmed to die out in the thymus. On the other hand, those T-cells with low tendency to react to self (receptors with low attractions), will survive and grow up to become mature T-lymphocytes. This is our immune system's protection process—we don't want T-cells to be self-reactive.

Once they've developed in the thymus, it's time for the T-cells to leave the cradle and face the abundance of world out there—in the lymph nodes and the spleen. This is where the real education begins for T-cells.

In the lymph nodes and the spleen, T-cells encounter a plethora of foreign antigens presented to them by other cells that patrol the body: macrophages, B-cells, dendritic cells. Once the TCRs come in contact with these foreign antigens, they rally and assume specialized responsibilities. So without a properly functioning TCR signal, there might be no immune response or the T-cell might turn against host body parts (self), which is known as *autoimmunity*.

A jacked-up immune response becomes a problem when a transplant or graft is given to a patient. In the treatment of cancer, bone marrow transplant (e.g., a stem-cell transplant such as what Jude received for her multiple myeloma) is an option. But after a transplant, serious rejections can happen, which can kill patients. In this way, the cure becomes a killer due to the incredible power of the T-cells. This is known as *graft-versus-host* disease.

When Naval Academy graduate and medical doctor Carl June came to train at the Fred Hutch Cancer Center in Seattle in 1983, he had a thought about graft-versus-host disease experiences: could T-cells taken from the patients themselves be tweaked to combat cancers? "The Hutch" was known as one of the premier institutions for bone marrow transplants in the world and a great place to learn from experts like E. Donnall Thomas, who conducted the first bone marrow transplant in the 1970s.

The U.S. Navy sent their man to the finest place to learn about the transplant of stem cells.

June was profoundly influenced by a simple dogma: to become fully activated, the T-cells needs two pushes—first, an antigen-specific push that comes from patrolling cells (antigen-presenting cells) in the lymph nodes and spleen; and second, an antigen-nonspecific push that comes from cell-to-cell contact. The difficulty was to demonstrate these T-cell activation processes in cell lines consistently.

I noted before in the BTK inhibitor journey that cell lines—which are cells resembling a given disease feature (i.e., models) cultured indefinitely in laboratories—aren't always faithful. They don't always yield similar results compared to experiments conducted on actual tumor cells derived from patients.

June determined that differences in T-cell behavior were owing to the variable level of incitement T-cells received in their milieu. In one of his early projects in Seattle, June showed that aside from the TCR itself, T-cells need encouragement from another receptor on their surface called CD28. This was the second push T-cells required for full activation mode. Recall from Tak Mak's analogy that he likened CD28 to the accelerator of T-cells.

June was onto something.

VIRUSES AS TOOLS

Spring 1986
Bethesda, Maryland

The world was learning about a new, deadly disease. A virus that was seen as a sporadic problem in predominantly African countries had started to spread in the United States. Yet it received a muted response from the government because of the stigma attached to HIV and the disease it caused. It was a so-called gay disease because homosexual men were the primary group affected by it. Drug abusers and racial minorities also suffered, and a swift response didn't result.

Inevitably, HIV started to appear in the mainstream media as the virus transmission became widespread. The *New York Times* front page on March 25, 1983, read, "Health Chief Calls AIDS Battle 'No.1 Priority.'" By fall 1985, President Ronald Regan had doubled down on the government's top priority: fighting HIV.

At this point, Carl June had to turn his research away from cancer to HIV/AIDS as the U.S. Navy prioritized infectious diseases. This suited June as viruses represent immunological elegance in taking over the cells they infect. In his case, HIV was appropriate because it targeted T-cells in humans.

June moved to Bethesda, Maryland, and his own research laboratory in 1986 to concentrate on HIV research.

Viruses need living cells to live in and allow them to multiply. HIV targets mainly T-cells that have a specific protein called CD4 on their surface. These are called *helper T-cells* because they help killer T-cells by alerting them to foreign antigens. By invading helper T-cells, HIV kills them. In HIV–infected patients, the number of helper T-cells gets too low, leading to the immune system's inability to process and alert foreign threats. That's why the disease is called *acquired immune deficiency syndrome*, or AIDS.

Back in Bethesda in his laboratory, June was working on obtaining T-cells from patients with late-stage HIV infections and growing them in culture. In this way he would have a pool of HIV in his laboratory that could be used later for more studies.

To June's surprise, his team failed to obtain HIV from the cultured T-cells.

Turns out they'd been using CD28 protein, which June had identified previously as a necessary stimulator for T-cells, as a means to continue growing the T-cells in culture. The CD28 protein ended up stopping HIV from entering the T-cells—an unexpected antiviral effect. Eventually June and Bruce Levine, his postdoctoral fellow, were able to grow T-cells in cultures by blocking the action of CD28 with antibodies.

The team now had liters of T-cells in flasks and ready to be used. This presented a great opportunity: because T-cell counts go too low in AIDS patients, these patients could be given laboratory-cultured T-cells as a treatment. Cell transfer was a proven method, championed by another trailblazing scientist and surgeon at the NIH.

Learning about June's success in culturing T-cells, a San Francisco–based biotech, Cell Genesys, reached out to June at this point. Their proposal was to create a *chimeric antigen receptor* (CAR) targeting HIV antigens and start a clinical trial for AIDS patients with these anti–HIV CAR–T-cells. If you look up the word *chimeric* or *chimera* in a dictionary, you'll come across a definition using words such as *fanciful* or *fantastic* or *imagination*. In Greek mythology, the chimera was a monster that was part lion, part goat, and part serpent.

In biology and genetics, *chimera* means a mosaic or hybrid of more than one tissue type. Thus a CAR is a a receptor that is not natural like TCR but has been engineered to help T-cells recognize antigens that they wouldn't normally recognize and act upon.

I like to compare CAR to the night-vision goggles that special ops use. Without this optical tool they wouldn't be able to see a threat in the dark, but with these goggles on, they can identify and neutralize threats. CARs are special goggles for the T-cells in our immune system. Armed with CARs, the T-cells can see and redirect their attack to a threat such as a virus like HIV or a cancer cell.

BEFORE CAR

Fall 1976
Bethesda, Maryland

Steven Rosenberg, recently appointed chief surgeon of the National Cancer Institute (NCI), was inspired by a paper published by Doris Morgan and colleagues in which the scientists described a growth factor (a secreted protein) from lymphocytes that could support the growth of T-cells in the laboratory. This growth factor was later dubbed *IL2* (interleukin 2) during an international conference in Interlaken, Switzerland. Interleukin promotes interactions between *leukocytes* (white blood cells).

In the 1970s there was no effective immunotherapy for cancer. Scientists doubted if human antigens expressed on cancer cells could at all evoke a meaningful immune response. George and Eva Klein, among others, had demonstrated an acquired immune response could prevent growth of transplanted tumors in mice, but would such a response ever be therapeutic? Could the human immune system be trained to develop a counteroffensive against cancers growing inside a human being?

Rosenberg found a way to produce IL2 from a culture of mixed lymphocytes in a medium in his laboratory. He had a plan to use T-cells for treatment and jotted down five steps to achieving this.

Rosenberg asked himself these questions:

1. Can I grow T-cells from both animals and humans in culture and maintain their killing activity?
2. Could these T-cells grown in culture be infused back to an animal and retain their killing abilities?
3. Can I find cells in animals and humans with cancer that can target cancer cells itself, obtain these cells, and grow them in the laboratory?
4. Can I infuse these cancer reactive T-cells back into animals that in turn can either slow down or eliminate tumors inside?
5. Could all this be done in humans?

I've read his commentary "A Journey in Science: Immersion in the Search for Effective Cancer Immunotherapies" in *Molecular Medicine* many times. I still get goosebumps when I read it.

These five questions or steps that Rosenberg wrote are a testament to his prowess as a researcher: simple questions, clear thinking, and a stepwise approach to solving a problem. I always suggest that my audience or students study this commentary. It's a fantastic journey indeed and offers a beautiful insight into the pioneer's work.

Rosenberg went on to identify a specific lymphocyte from the supporting structure of a tumor known as the *tumor stroma*. In the stroma there are non-tumor immune cells, membranes, blood vessels and connective tissue. In this microscopic milieu, specific immune cells fight back against cancer cells, which are known as *tumor infiltrating lymphocytes* (TIL). These cancer-reactive TIL plus IL2 (which promotes T-cell growth) prove a powerful combo against cancer. TIL collected and grown from the stroma of a skin cancer called *melanoma* and boosted by IL2 regresses this cancer in humans successfully.

Cell therapy was no longer a pipe dream.

At the same time, Zelig Eshhar at the Weizmann Institute in Israel was working on the idea that the natural T-cell functions might be tweaked to make them recognize tumor cells. He proceeded to replace part of the TCR with antibody parts. We've seen before that antibodies have a *Y*-shaped structure with variable and constant regions. In his laboratory, Eshhar and his colleagues combined the variable region of an antibody with the constant region of TCR in 1989, creating a *chimeric receptor* on the surface of the T-cells that could detect and antagonize target cells that had the antigen of interest.

This was a genius approach. Rosenberg saw the promise of application of such a chimeric receptor for treating cancer. He invited Eshhar to work as a visiting scientist in his laboratory in Bethesda. The genetically modified TIL (CAR–T) proved safe to use in humans.

1986
Boston

Michael Sadelain, a French-born, self-proclaimed geeky medical graduate, was intrigued by Rosenberg's work in TIL. It was exhilarating to see T-cells collected from the tumor boroughs turn against the tumor itself, but Sadelain felt that relying on specific T-cells like TIL to fight cancer cells was too unpredictable. A more straightforward route would be to engineer primary T-cells (not TIL) to train them as tumor killers. This required genetic manipulation of T-cells—something that was considered mystic and too ambitious at the time.

Sadelain got into Massachusetts Institute of Technology to study gene transfer biology. His official project was to study a rare blood disorder, not cancer, but Sadelain secretly kept working on his plan to install genes in the T-cells. He found the process incredibly difficult. Others had also tried to use a virus to introduce genes inside a cell with little success.

As Sadelain's efforts of gene transfer seemed to have reached a dead-end, HIV dominated national headlines.

In Bethesda, Cell Genesys and June teamed up for a CAR–T clinical trial for AIDS.

Using Bruce Levine's T-cell culture system, June's lab and Cell Genesys showed successfully that CAR–T improved the immune functions of patients with HIV infections in a first-in-human Phase 1 trial.

Boosted by the success of the initial trial, Cell Genesys announced a Phase 2 trial of CAR–T-cell therapy for AIDS. When the news broke about this potential new AIDS treatment in 1997, Cell Genesys shares soared a whopping 43 percent overnight from $3.65 to $12.12.

However, the Phase 2 trial of anti–HIV CAR would prove to be a disappointment. The study found no difference in HIV virus accumulation in the blood and GI tract of those participants who received the modified T-cells as a treatment. On the other hand, antiviral drugs taken regularly showed great benefit in treating HIV, driving viral load to an undetectable level—a much easier option compared to engineered T-cell therapy. Soon the interest in anti–HIV CAR waned.

As for June, it was time to move on.

He has served in the U.S. Navy for twelve years, fulfilling his obligation to the Navy for supporting his education and training. Retiring from the Navy, he moved to Philadelphia and to the University of Pennsylvania along with other key team members, including Levine, in 1996.

June planned to focus on CAR–T treatment for cancer exclusively.

Chapter 14

CAR versus Cancer

A GAS PEDAL FOR CAR

The first generation of engineered T-cells was tested in trials in patients with ovarian and renal cell cancers. For ovarian cancer patients who had metastatic disease, the CAR–T-cells targeted a protein associated with ovarian cancer cells. Among the fourteen patients treated with this CAR–T, no one responded. The tumor burden did not decrease.

Similarly, another first-generation CAR targeted a sugar-protein molecule present on the surface of renal cell cancers, and three patients with advanced renal cell carcinoma received the treatment. Again the patients' disease progressed within one hundred days with no control of the spreading cancers.

These two scientific reports on early CAR–T endeavors, published within six months of each other in 2006, confirmed that CAR–T could be given safely to humans but that these cells were not able to survive in sufficient numbers inside the body long term. Once infused, the CAR–T-cells were detected in the blood of the patients for the first few days but their numbers quickly dwindled, and by the end of one month, CAR–T-cells were barely detectable in the blood.

This was a disappointment. What good is a living cell treatment if the cells get too tired too soon and perish? The T-cells need to persist, to stay focused on the enemy. They needed periodic boosting so that they wouldn't get exhausted. In other words, our CAR needed a gas pedal: a source of drive or acceleration that would keep pushing the CAR–T-cells, encouraging them to multiply and keep the fight on.

June had already identified CD28 as a co-stimulator of T-cells, something that kept the cells invigorated. If CD28 acted as the motivator for natural T-cell receptors, perhaps the same CD28 could also be a stimulator for an engineered receptor like CAR?

In Boston, Sadelain finally succeeded in introducing genes in T-cells. He moved to his own lab at the Memorial Sloan Kettering in New York in 1994 and decided to focus on T-cell engineering exclusively. His fondness for seemingly esoteric tasks in cell engineering ended up solving the problem of lack of a drive for the first-generation CARs.

Sadelain engineered a gas pedal for T-cells by adding CD28 to the CAR structure. His CAR had CD19 as the primary target and CD28 as its accelerator. Because CD19 is a common protein expressed by the B-cells (including leukemia and lymphoma cells) it could be an effective treatment against aggressive blood cancers.

But the CD19 CAR–T needed to pass the real test: a clinical trial with B-cell tumors. The race for a successful cell treatment was heating up.

A THREE-WAY RACE

In Philadelphia, June was entrenched in designing and experimenting with T-cells in his laboratory at the University of Pennsylvania. He was after an effective, individualized treatment for cancer that would be groundbreaking. There were several groups working on CAR–T projects at the same time—all of them in academic centers. The pharmaceutical industry had its doubts about tailoring treatments to individuals.

In 2003, June heard a scientific presentation by a pediatric cancer specialist, Dario Campana of St. Jude's Research Hospital.

He was intrigued.

Campana and his colleague Chihaya Imai had successfully designed a CD19 CAR–T with a less-known booster of the T-cells—a different gas pedal called 4–1BB. Seeing the promise, June requested a sample of Campana's CAR–T. Through an agreement between the University of Pennsylvania and St. Jude's, he got hold of the genetic blueprint of Campana's CAR.

In New York, Sadelain had designed his CD19–targeting CAR–T with CD28 as the driver. In contrast, June had a CD19 CAR–T with 4–1BB as its accelerator (making use of Campana's CAR blueprint). About the same time, a young blood-cancer fellow at the NCI, James Kochenderfer, approached Rosenberg about constructing a CD19 CAR–T treatment for lymphoma patients at the NCI.

All wanted the same thing: to take their cell treatment to the clinic.

In those early days of cell treatment research, funding was limited. It was hard to get money for a clinical trial with CAR–T. Like Campana, Sadelain shared his CAR–T design with prominent researchers in the field, including June and Rosenberg.

Despite the collaborative nature of the work, it was a three-way race for a remedy at this point: June's team in Philadelphia, Sadelain's group at Memorial Sloan Kettering in New York, and Rosenberg's gang at the NCI in Maryland. All were competing to publish the results of their separate trials of cell treatments in human cancers.

Rosenberg and Kochenderfer's CAR passed the checkered flag first.

In fall 2010 while in training, I was tasked with reviewing new progress in immunotherapy for our fellows and visiting scientists. Kochenderfer's first paper with the clinical results of an anti–CD19 CAR–T had just been published, and I ended up reviewing it.

The science was simply brilliant.

Using a CD19 CAR–T, the NCI group treated a patient who had stage 4 follicular lymphoma. The lymphoma had spread through his body despite multiple treatments. All of the patient's lymph nodes were packed with lymphoma cells (cancerous B-cells), and the patient had lost ten pounds in a matter of days. The signs of cancer progression were unmistakable.

Before treatment, the medical team at the NCI did a bone marrow biopsy that revealed about 35 percent of the cells in the patient's marrow expressed CD19 protein. They collected cells from his blood to prepare the CAR, and after the T-cells had been engineered, he was given a first dose of 100 million CAR–T-cells. The following day another 300 million CAR–T-cells were given intravenously.

"They gave millions of living cells to a patient?" The reaction from my small audience was one of bewilderment. They couldn't believe that this had taken place already. The patient was discharged from the hospital eleven days after receiving the second dose of cells and he went back to work full time.

It was astounding.

We were familiar with traditional medicine. A pill taken orally or an injection was something we were used to. But this was a whole new paradigm, a living drug exclusively for use by the very same person whose cells were used to create it, it sounded like science fiction.

"Can we call it a drug or medication?" quipped a fellow doctor.

I hesitated. I didn't know if we could call cell treatment a drug. Unlike small molecules, it was not chemically synthesized. And it was not a product secreted by living cells either, like an antibody. It was the cell itself genetically modified to become a treatment.

"It's certainly a remedy," I offered. "And a great one for that matter."

"Can you please explain how it's made?" a fellow asked.

"I'll try," I said.

"The first step is to collect T-cells from the patient. Through an apheresis machine, the patient's blood cells are separated from liquid plasma. In this way, T-cells are isolated and the rest of the blood is returned to the patient. The harvested T-cells are then transferred to a laboratory."

I paused. Judging from the facial expressions of my audience, they were hooked.

"The cool stuff begins in the lab. Using a virus for the delivery job, a specific genetic material is introduced to the T-cells. Lentivirus or retroviruses are good at accepting genes and carrying them within. In the culture, the virus carrying the CAR genes attaches itself to the T-cells' surface and delivers its genetic material into the T-cells. Since viruses are experts in taking over the cells they infect, the viral genetic material, including the CAR genes, becomes a part of the host T-cells' genome. The genome goes on to produce the protein—chimeric antigen receptor—on the surface of the T-cells. These T-cells then multiply in their billions in the culture—all armed with that chimeric antigen receptor—on the surface of the T-cells. These T-cells multiply in their billions in the culture. They are then kept frozen, ready to be infused back to the patient."

Having explained the procedure, I moved on to the clinical response seen in this patient.

The scans of the patient revealed a partial remission of lymphoma after CAR–T treatment—a remarkable result given the lymphoma was aggressive. CAR–T kept the cancer at bay for a straight thirty-two weeks. This was the first time a CAR–T treatment had achieved remission of a cancer unlike the first generation that failed to reduce tumor burden in ovarian and renal cancers. The patient's blood did not show B-cells for almost four months, which meant millions of CAR–T-cells had acted on the B-cells that exhibited CD19 (both cancerous B-cells and normal B-cells) and eradicated them from the blood. Even in the patient's bone marrow, the B-cells with CD19 were wiped out for a prolonged period.

Fascinating!

"This is just one patient, Mak," one of my colleagues said, trying to throw in a bit of a caution. Often the very first patient doesn't represent what may happen afterward.

"True. The first patient, but not the last—I have no doubt in my mind!" I said.

From these results it was clear that the CAR–T-cells not only survived for a prolonged period inside the patient's body but also selectively obliterated CD19–expressing B-cells with amazing efficiency.

True to my anticipation, ten months after the publication of Rosenberg's first patient's data with CAR–T, June's team at the University of Pennsylvania published their own patient's findings, this time with a CD19 CAR coupled to a 4–1BB gas pedal.

Doug Olsen had a case of chronic lymphocytic leukemia (CLL) that kept coming back despite rounds of chemo. In spring 2010 his oncologist told him that the cancer had become resistant to all standard treatments. Half of his bone marrow was occupied by cancer cells—an ominous sign.

Olsen's oncologist, David Porter of the University of Pennsylvania, was working with June and had a CAR–T trial in mind for his patient. When he offered Olsen the option, Olsen read through the documents to understand the process of getting CAR–T.

"It was a bit strange to read the literature and descriptions of how this process had worked in mice and to realize that the next set of experiments was going to be with humans—specifically me!" While it was unnerving, Olsen did not hesitate to accept the risk. "I was definitely in—with no reservations." It was a chance for him to fight back, to not just be a victim of cancer.

Olsen was the second patient to enroll in the CAR–T trial at the University of Pennsylvania. Over three consecutive days he received about 14 million CAR–T-cells—a lot fewer compared to the earlier CAR–T trials. The CAR–T-cells multiplied in Olsen's body quickly and by six months the tiny soldiers had taken back the bone marrow from the grasp of the tyrannical leukemia cells. The CAR–T-cells multiplied about one thousand times on their way to crush leukemia.

A beaming David Porter walked into Olsen's room one morning and announced, "Hot off the press, 18 percent of your white cells are now CAR–T-cells—the transduced cells are reproducing and killing cancer cells!"

Three weeks later, Porter confirmed that they couldn't find a single cancer cell in Olsen's body: his bone marrow was completely free of leukemia cells. Olsen decided to declare victory with a big "f—k you" to cancer and drove straight to a boat show in Annapolis with his wife.

The couple bought a sailboat.

In his note to the University of Pennsylvania team—David Porter, Carl June, and others—Olsen wrote,

Dear Penn Team:

My New Year's wish for all of you is that, every single day, you take a moment to think about and take pride in what you are involved in and what you have accomplished. I am just one life but I touch others. I am convinced that you have found the key that will save thousands who touch millions.

Warm regards, Doug

Hope once again proved to be the best companion for a cancer patient. For Doug Olsen, it was founded on a groundbreaking treatment for cancer. His story and the results of June's CAR–T was transformational.

The tide had turned.

For years, big pharma hesitated to invest in personalized treatment. It's much easier to create treatments which can be manufactured for a large group of patients who have the same cancer. Cell treatments must be made from individual patients, which is a process difficult to scale up. But seeing the extraordinary ability of cell treatment to induce cancer remissions, pharma started to invest in the opportunity. CAR–T was well and truly on the road to becoming an incredible remedy.

Chapter 15

The Wrath of T

I'd like to avenge myself at once, as you advise, but I've always feared your wrath and shied away.

—Homer

THE FIRST CHILD

Of all blood cancers, I found it the most difficult to witness the impact of *acute lymphoblastic leukemia* (ALL). It is the most common cancer in children and adolescents, and about two-thirds of the four thousand patients that are diagnosed each year in the United States are children. At an age when they are supposed to spend time learning and having fun, life turns into seemingly endless episodes of visiting hospitals and being treated.

When will I get better? When will all this go away and I can go back to school? It's hard to answer these questions from children who don't know the kind of enemy they're battling.

Yet there is one thing about children.

My friend, blood cancer expert Simon Rule would say, "You give them any treatment—chemo, transplant, targeted drugs—they can tolerate it!"

It's true. At a younger age the body tends to suffer fewer and less-serious side effects from potent anti-cancer agents. Of course this is not a universal truth and there are different consequences with different treatment approaches. Among children, chemo treatment has great outcomes for ALL. During the 1990s, about 70 to 80 percent of children would survive for five years with ALL. The likelihood of survival is beyond 90 percent now—a testament that children tolerate and respond better to intensive treatment.

Sadly though, not all can be cured with chemo regimens.

Spring 2010
Philipsburg, Pennsylvania

Tom and Kari Whitehead's worst nightmare had come true. Their daughter, five-year-old Emily, had just finished preschool and her parents were looking forward to a summer full of fun watching her play hard, go camping with the family, or splashing at the pool.

Instead, Emily fell sick.

What started with seemingly trivial bruises was not any minor illness; she had acute leukemia, a dreadful cancer of the B-cells.

The world came crashing down for Tom and Kari. An electric power line-man by trade, Tom received the devastating news from his wife while at work. He had to rush to the emergency room at the regional hospital to be with Emily. Kari confirmed the diagnosis over the phone later.

At diagnosis, Emily's cancer was what the experts call *standard risk acute lymphoblastic leukemia*. In this disease, the B-type white cells grow in an uncontrolled fashion in the patient's bone marrow. Because these tumor cells (or *blast-cell*s, an immature B-cell) grow fast, they displace mature, normal blood cells such as red blood cells and platelets in the marrow. Soon the aggressive takeover of the marrow is complete and the leukemia cells spill over into the blood. With no additional genetic risk identified, Emily's B-type ALL would be treated with chemotherapy. As noted, chemo is very effective for children with this cancer, and this is typically good news for the family given the circumstances.

Emily started treatment with chemotherapy at Hershey Medical Center, about one hundred twenty miles southeast from her home in Philipsburg. In this case, her ALL relapsed after the chemotherapy and twenty six months from the diagnosis. Her bone marrow became crowded with leukemia cells once again—a sure sign that the cancer had returned.

When ALL relapses from initial chemotherapy, the outcome's usually not good. At their first relapse, only about 50 percent of children with ALL will survive five years. With second or third relapses, there'll be only a few months of survival.

There are a myriad of indicators for poor response to treatment in ALL: faster progression from diagnosis to relapse, reemergence of leukemia in the bone marrow, and, in general, less effective response to initial treatment are a few known factors. Several of these factors were present in Emily's case.

Emily had a relatively faster progression to relapse so she was no longer a standard-risk ALL. At this point, stem-cell transplant was an option, but the patient needs to be eligible to start such a complex procedure.

Emily's mother Kari wanted to have a second opinion from a larger hospital, and Children's Hospital of Philadelphia (CHOP) was on her list. The family consulted the oncologist at CHOP. It was still a matter of getting Emily ready for the transplant, and the family thought they would keep her close to home, at Hershey, where she was getting the treatment, instead of Philadelphia.

Getting a bone marrow transplant isn't a straightforward business. The ALL needs to be controlled to a remission by chemo, the cells for the transplant need to be obtained from a donor, and the blood counts must be within an acceptable range. After an agonizing period of uncertainty, Emily's parents were informed by the doctors that Emily had relapsed again and was no longer eligible for a transplant.

There was no standard of care after the second relapse. They could try another intensive chemo combo to tame the exploding tumor but that was all.

Kari and Tom thought to defer. Not another round of chemo.

They decided to take Emily to CHOP. They would seek a newer, experimental treatment in a last effort to save their daughter's life.

At CHOP an immunotherapy trial for ALL was about to begin with CD19 CAR–T as designed by Carl June's team. At the time Tom and Kari brought Emily after her second relapse, the trial did not have clearance from the FDA to start enrolling. This clearance is known as "The study may proceed" letter: a final decision from the FDA before a clinical trial can begin.

There being no other alternative, another round of chemo had to be started back in Hershey but the cancer was stubborn and would not go into remission. At this point all options pretty much ran out for the desperate couple.

Emily's days were numbered. Tom and Kari were told to go home where Emily could spend the last few days of her life in comfort. Refusing the suggestion, Tom reached out to CHOP. One last call for help, one last hope?

Tom was informed by the CHOP team that the CD19 CAR–T trial was now enrolling, and what's more, Emily could be the first to receive the cell treatment.

Wasting no time, the family headed to Philadelphia.

We've seen that in order for the body's T-cells to unleash their fury against cancer cells, they must first recognize the cells as foreign. Cancer cells are our own cells so it's not natural for T-cells to turn against self automatically. But if we can teach the T-cells to see a cancer cell via an identification marker on that cell, they can attack it.

A protein called CD19, present on the surface of the B-cells (leukemia is a B-cell cancer) provides that target identity for B-cell cancers. This protein is present on B-cells from an early time in its development, all the way until B-cells become plasma cells.

Armed with a CAR—a kind of talon on the surface of T-cells that has an antibody part (a chimera) that recognizes CD19 protein—the CAR–T-cells have little problem seeing cancer cells. Using this concept, Carl June's

CD19 CAR–T with 4–1BB had already delivered Doug Olsen his remission from chronic leukemia.

But Olsen was an adult, and his cancer was different from Emily's. Could this CAR–T put up a fight against ALL cells, which are masters in evading immune attack? This had never been tested in a child before. For Emily's parents and her oncologist Stephan Grupp, CAR–T was the only option. Grupp designed a clinical trial teaming up with Carl June, Bruce Levine, and David Porter—a dream team if you're getting CAR–T.

Maybe, just maybe, Emily was at the right place at the right time.

A CYTOKINE STORM

April 17, 2012
Philadelphia, Pennsylvania

After a harrowing two years battling aggressive ALL, Emily's own T-cells were collected from her and sent to a production lab at the University of Pennsylvania where an anti–CD19 CAR was genetically combined with the T-cells. The cells were grown in their millions, and this army of CAR–T-cells was infused into Emily one month after the original T-cells had been collected.

Tom, Emily, and Kari wrote about the pivotal moment in their book *Praying for Emily*.

"Dr. Grupp himself attached the syringe full of CAR–T-cells to her port. We had so much hope riding on these cells." To her dad's plea for his baby girl to be strong, a sleepy Emily replied, "I will fight as hard as I can, Daddy."

For the first time in history, a child was receiving treatment with genetically modified cells.

On the following day, Emily received her second round of CAR–T infusion, having received only a fraction of the cells on the first day of treatment.

When the CAR–T-cells were infused, they immediately engaged with the tumor cells floating in Emily's blood that bore the target of interest. In this case, the CD19 protein was the target. The higher the number of leukemia cells with CD19, the more intense the battle between CAR and cancer.

It would be naive to think that such a major war between T-cells and cancer cells won't have any repercussions. The calm of the infusion is the eye of the storm. Right after the hush, a chaotic whirlwind's waiting. Indeed, it is expected that the patient will get sick after the infusion of CAR–T-cells—this is a sign that the treatment is actually working.

As the CAR–T-cells are growing and dividing in the body, they instigate an acute reaction, which is the expected response of the immune system. This is known as *cytokine release syndrome* (CRS). Cytokines are small proteins, a form of chemical messenger of the immune system that tells immune cells to grow rapidly and fight an infection.

CRS turned out to be a category 5 hurricane for Emily.

Her condition deteriorated dramatically after the remaining CAR–T had been infused on the third day of cell treatment. On the afternoon of the infusion, she spiked a fever with temperatures hovering around 105 degrees. She started vomiting and her blood pressure dropped to a dangerously low level. Emily was struggling to breathe. She became disoriented and progressively more listless. Because the oxygen levels in her blood were plummeting, she was transferred to the intensive care unit.

On April 23, five days after the first CAR–T infusion, Emily was put on a ventilator and into an induced coma.

While this response to CAR–T was an expected event, the seriousness of Emily's CRS surprised her medical team. From their experiences with adult patients like Doug Olsen, the team knew if they could ride the CRS, the CAR–T-cells might take control over the cancer. But how long would a five-year old weather the storm?

Emily was fading away.

Carl June, the architect behind the CD19 CAR, and Stephan Grupp, the pediatric oncologist who was in charge of Emily's care at CHOP, were certain she would not survive the night. Tom and Kari were informed of their daughter's condition: she had a one-in-a-thousand chance of making it to sunrise. Both her kidneys and lungs were failing—five-year-old Emily was teetering on the edge of death.

CRS is a clinical diagnosis. There isn't a cell count or lab marker that can irrefutably say someone has CRS. First and foremost, the patient must have fever that is not due to any other cause (like a bacterial or viral illness). Second, the patient could have low blood pressure (*hypotension*), and third, the patient's oxygen levels in the blood (*hypoxia*) could be low. Using fever, hypotension, and hypoxia, the severity of the CRS is determined. Although the diagnosis is clinical, we can monitor for different cytokine levels in the blood and correlate them with the CRS.

As Emily walked the fine line between life and death at the pediatric ICU of CHOP, her bloodwork revealed high levels of a particular cytokine, IL6 (interleukin 6). It was several-thousandfold higher than normal. Like other interleukins, IL6 is secreted by activated immune cells, including T-cells. But no one knew for sure what role IL6 was playing in Emily's situation.

June seized on the laboratory results. He knew from experience there was an antibody available that could antagonize IL6 known as *tocilizumab* or simply *toci*. At the time, tocilizumab was already approved for the treatment of arthritis in children. It was not a recommended remedy for a serious cytokine storm. But what else could they do to save this young soul?

June convinced Grupp and Emily's ICU team to treat her with tocilizumab. She had hours left before falling into the abyss from which no one could return—the clock was ticking.

Out of desperation the medical team gave Emily a dose of tocilizumab in the evening. It was logical to fight the CRS with this antibody given her IL6 levels.

It marked a spectacular turnaround for Emily.

Within hours, Emily's fever had subsided. Her low blood pressure quickly reversed to rapidly climbing blood pressure. Grupp had to immediately take Emily off the blood-pressure-boosting drugs she was receiving to protect her from possible hemorrhage. By sunrise, Emily was out of her CRS.

On May 5, eighteen days since she'd first received CAR–T, Emily was off the ventilator. She could breathe on her own once again. Bone marrow aspirate was drawn and CT scans were completed to review her disease status in the following days.

Later, Grupp gave the amazing news to Tom over the phone: the CAR–T had worked brilliantly. There were now no leukemia cells in Emily's bone marrow. What high intensity chemotherapy couldn't do for more than two years, CAR–T did in less than a month. It was an extraordinary feat for modern-day medicine.

The first child in the world to receive CAR–T was finally free from acute lymphoblastic leukemia, defying death. Emily became a poster child for CAR–T treatment. Her story was published all over: newspapers, magazines. Together with this cutting-edge treatment, the five-year-old girl from Philipsburg became a sensational figure who showed the world what could be accomplished with cell therapy.

CARS ON THE MOVE

Spring 2014
Houston, Texas

About 1,500 miles southwest of Philadelphia, down in Houston at the MD Anderson Cancer Center, I was busy crafting clinical projects for the B-cell Lymphoma Moonshots Program. Although not directly involved with the cell therapy unit, I was intrigued by CAR–T. As a super-specialized cancer research center, we had a great array of scientific and technical skills at the institution. Our vast patient population gave us a unique advantage in conducting CAR–T trials: people from all over the country sought care here. At the time, cell therapy was mainly done at the Stem-Cell Transplant Department at the MD Anderson Cancer Center.

In February 2014, Michael Sadelain's group at Memorial Sloan Kettering Cancer Center in New York published their CD19 CAR–T trial findings. Sixteen adult patients with ALL had received CAR–T and 88 percent had achieved complete remissions of their disease—doubly better than any chemotherapy.

Toward the end of that year the University of Pennsylvania–based group led by Carl June published their CAR–T trial initial findings. In their paper in the *New England Journal of Medicine*, June's team reported data from thirty patients, both adults and children, the very first of whom was Emily Whitehead. Of these, twenty-seven patients (90 percent) achieved complete remission of their ALL. Every patient developed a CRS, and some were severe. Like Emily they needed tocilizumab treatment to recover from their CRS.

To give you a sense of how competitive the different CAR teams were, just two days before June's group published, Rosenberg's NCI group published their own findings online in the *Lancet*. They reported a 70 percent complete response rate for their CD19 CAR–T in children and young adults—slightly lower than the other two groups. Their trial included lymphoma patients in addition to ALL patients unlike the other two trials.

No other period in my career has been as exciting as the years from 2012 to 2014. T-cell therapy was a revolutionary treatment. No other cancer treatment was producing this kind of response. A lot of us started talking about cures for advanced blood cancers without chemotherapy, something that would've been extraordinary just a couple of years before.

But while the response to CAR–T treatment was superb, another side effect other than CRS was emerging as a serious concern.

After receiving CAR–T, a few patients became confused and disoriented and were unable to speak. In some cases, the patients were stupefied. Some became delirious, even hallucinatory. The NCI group noted abnormal brain MRI findings in one patient while Sadelain's group reported three of their patients had needed ventilator support following neurological symptoms. Clearly the CAR–T treatment was causing these problems with the central nervous system.

But why did the CAR–T-cells go to the brain? This wasn't clear to the experts.

Although this apparent toxicity was concerning, the commercial development of CAR–T therapies for cancer exploded. Several pharma companies raced to acquire and develop cell treatments. Novartis partnered with the University of Pennsylvania team; Santa Monica–based Kite Pharma licensed the NCI CAR–T. Michael Sadelain cofounded Juno Therapeutics in collaboration with the Fred Hutch Cancer Research Center, Memorial Sloan

Kettering, and the Seattle Children's Research Institute with the intention of commercializing his CAR–T therapy.

At this point, my colleague at the MD Anderson Cancer Center, lymphoma expert Sattva Neelapu, was invited by Kite founder and CEO Arie Belldegrun to join a scientific session to discuss findings from the NCI trial. Kite needed to better understand the implications of brain toxicity having secured the CAR–T deal with NCI.

Neelapu flew to Bethesda to participate in the discussion. He wanted to bring the CAR–T trial to our department in Houston.

RACE TO THE FINISH LINE

Fall 2014
National Cancer Institute
Bethesda, Maryland

In a day-long meeting at NCI, Neelapu met the heavyweights of the cell therapy world and several lymphoma experts: Rosenberg, Kochenderfer, Levy, Belldegrun, and David Chang (Kite CMO), among others. The scientists discussed how best to manage neurotoxicity.

Although Kite had licensed the CAR–T technology from NCI, they hadn't transferred the manufacturing of it to their own facility yet. The prepwork was underway. Kite's nearly 20,000 square-foot cell-therapy manufacturing facility in Santa Monica was set to begin operations soon.

The discussion at NCI revolved around starting a Phase 1 trial with CD19 CAR–T and moving on to a Phase 2 trial that would hopefully lead to FDA approval. The primary target was aggressive lymphomas such as *diffuse, large B-cell lymphoma* (DLBCL).

Neelapu agreed to participate in the Phase 1 study. It was exciting for all of us to bring the CAR–T trial to Houston. The Phase 1 trial was named ZUMA-1.

Our department started enrolling patients in the trial in 2015. Kite announced an expanded cell-therapy collaboration deal with NCI, and the company's quarterly payment to the nation's principal agency on cancer increased to an eye-popping $750,000.

Despite the excitement, Neelapu had some concerns.

"I knew it would be very different from what we'd been doing so far," he said to me. "All those early trial experiences, especially from Kochenderfer, pointed to the possibility of serious toxicities with CAR–T."

Solid team coordination is crucial for the success of any therapy. CAR–T treatment is a perfect example of this. The treatment process with CAR–T is complex and requires active participation from different healthcare providers, including doctors, nurses, coordinators, nurse educators, apheresis nurses, inpatient nurses, outpatient nurses, case managers, and social workers.

Neelapu wanted to make sure our nurses were onboard and trained before this new trial came to Houston. Our nursing team was super experienced in administering chemotherapy and other targeted treatments and they knew how to monitor cancer patients very well but monitoring CAR–T treated patients required a different skill set.

Over the following weeks, our nurses were trained on what to look for in a patient who had received CAR–T and how to monitor for signs of an acute reaction such as CRS or neurotoxicity. Education sessions were run at different times of the day with Neelapu on call 24/7.

On the day the first patient, an adult male of seventy years, was to be infused with the CD19 CAR–T at the MD Anderson Cancer Center, a crowd of about twenty people were gathered in the room.

"The infusion was so simple," Neelapu chuckled, as he recalled the moment.

Like Emily, Doug, and others, the infusion itself was very easy and calm. Nothing out of the ordinary. After the infusion, our senior-most nurses were assigned to monitor the patient day and night. Neelapu went to the bedside several times a day to make sure all was okay. At first the patient developed a mild fever but that was expected and managed easily.

On day 5 though, things took a sinister turn: the patient developed neurotoxicity.

"I knew from other's descriptions about what could happen," Neelapu said. "But seeing it live for the first time was scary."

The patient who received CAR–T was a nuclear physicist. He had acknowledged the risks involved before starting the treatment. His wife was very supportive as well—she actively participated in all the educational sessions as he was being prepared for the first infusion.

The neurotoxicity began with mild confusion. Then the patient became totally *aphasic*, losing the ability to understand and communicate with words. This is typically seen after a brain injury such as stroke when the area of the human brain that processes language is damaged.

The patient would randomly sit up in bed, looking around the room vacantly. It was two-way aphasia: he didn't understand what was being said and couldn't express how he was feeling. He didn't know where he was. For a physicist who relied on razor sharp analytical and communication skills to explain the features of the center of atoms, this was akin to death. He had lost his most critical faculty: the ability to express himself.

This went on for three days.

"His tumor was only about an inch in diameter. I honestly thought he would sail through," Neelapu said. Investigators don't typically enroll a high-risk patient when they are treating with an experimental remedy for the first time. Although this patient had DLBCL, that cancer often produces a substantially large tumor—more than three inches in diameter is not uncommon. Generally the larger the tumor before starting treatment, the higher the risk for acute treatment-associated reactions such as CRS or neurotoxicity. Nonetheless, neurotoxicity developed.

On the fourth day of neurotoxicity, the patient turned around.

He started to communicate and to respond to instructions from nurses. By the end of the first week, the physicist had completely recovered from neurological dysfunction.

"This was a huge moment," a relieved Neelapu said. He could feel the burden lifted off his chest. He was not prepared to say goodbye to the patient and dreaded explaining to the patient's wife what had happened. But the real cause for celebration came at the end of the month. The patient's scans came back showing he was free from lymphoma—a complete response from the CD19 CAR–T!

I was not surprised.

This had started to become a consistent story: patients with aggressive blood cancers, having failed several previous treatments, were becoming disease-free after CAR–T treatment. Of the first six patients treated in the Phase 1 trial in our institution, five had complete remission of their lymphoma while the sixth patient achieved 99 percent clearance of the tumor (a small speck of active cancer cells prevented a complete response). This patient ended up being a very good partial responder.

Having shown CAR–T's power in wiping out CD19–exhibiting B-cells in ALL, early data with CAR–T were already showing similar unprecedented results in a second blood cancer, DLBCL.

Kite knew they had an amazing product on their hands. While ZUMA-1 was ongoing, in November of 2015 they started a Phase 2 trial for MCL patients who had relapsed from up to five prior treatments including a BTK inhibitor drug. Michael Wang was set to lead this trial in MCL, dubbed ZUMA-2.

As more data emerged from CAR–T trials in leukemia and lymphoma, we started getting a better picture of the workings behind CRS and neurotoxicity.

CRS is still the most common immediate adverse effect of CAR–T treatment. Once inside the body, CAR–T-cells recognize tumor cells, and assuming combat mode, they release cytokines. What's more interesting is that the CAR–T-cells also switch on bystander cells in the tumor milieu, like macrophages, that also take part in the cytokine storm. Emily's situation taught us the importance of monitoring IL6 following CAR–T infusion. But there are others—IL8, IL10, IFN-g, MCP-1—and these proteins secreted by the immune cells also play their part in the cytokine response. Tocilizumab, which blocks IL6 alone, might not be sufficient for treating CRS so experts resort to steroids for suppressing total immune function in addition to tocilizumab as needed.

Neurotoxicity is a whole different story. It may or may not happen alongside CRS. Sadelain's group had three patients with serious neurotoxicity in their early trial. The team tested *cerebrospinal fluid* (CSF; a colorless liquid found in the tissue that surrounds the brain and spinal cord) from these three patients and confirmed the presence of CAR–T-cells in the fluid.

This was surprising. CAR–T-cells were supposed to seek out cells that express the target—CD19—on their surface. Since there were no CD19–bearing tumor or immune cells in the brain of these patients, why would the CARs travel all the way up there and park? The *blood brain barrier*, a form of biological blockade formed by the cells that line the blood vessels of our central nervous system, strictly regulates what enters the brain. Paul Ehrlich couldn't stain brain cells with dye because of the blood brain barrier.

What could have caused the CAR–T to break through? Perhaps the cells lining the blood vessels in the brain were disturbed and became leaky due to the action of the CAR–T-cells?

There were suggestions that the brain's own immune cells like macrophages were being activated by CAR–T-cells and playing a part in the neurotoxicity. Serious neurological problems continued to be managed with high-dose steroids in the absence of a definite understanding of what was going on. To this day, neurotoxicity is a very big concern for CD19–targeting therapies. As for the clinic, we kept learning, and we kept treating one patient at a time with CAR–T.

December 2016
San Diego, California

Neelapu took the stage to present findings from Kite's CD19 CAR–T trial for DLBCL patients at the ASH annual meeting.

Incredibly, the ZUMA-1 trial had already treated one hundred patients by December. It was amazing that a novel mode of treatment should be adopted so quickly, and I knew first hand how dedicated the entire crew was at our center. Doctors, scientists, nurses, coordinators, technicians, patients, caregivers—everyone learned and understood their role. An almost sacred sense of heralding in a new generation of treatment inspired the team. You could see that dedication on the faces of the nurses and supporting staff and feel their positive energy.

"We are doing something big!" they would say with pride.

The data from the trial was made available as a late-breaking abstract, meaning it was compelling new information that had emerged after the usual deadline for submission to the ASH annual event.

Neelapu was the right man for the job.

One thing that impresses me most about him: he is never flustered. He soaks it all in—comments, suggestions, rebuttal—and still seems to have the time to process that information and provide a thoughtful answer. He's not a dynamic presenter (like Wang), but whenever he's up there on the podium, you can expect great scientific insights.

It was the last presentation on the last day of the conference.

Neelapu presented findings from fifty-one patients. About half of them (47 percent) had complete remission of their DLBCL and were monitored for at least three months after treatment. A good number of the patients in the study had had very aggressive, bulky tumors. Even so, they responded well.

But experts were divided on the matter.

"There was a lot of skepticism from some of the lymphoma experts," Neelapu said. "People had their doubts."

At that point in 2016, the follow-up period after treatment was only about six months so the team did not present data on how long the patients survived without progression of their cancers. This period of *progression-free survival* (PFS) is a common proxy for survival benefit. Without this data, it's hard to convince critics of the benefit of a new treatment.

Others thought the patients should have been allowed to receive other forms of immediate treatment (known as *bridging therapy*) while their CAR–T was being produced. By design, ZUMA-1 didn't allow bridging therapy. All of these meant there could have been *selection bias*, which would mean Neelapu and colleagues might have indirectly chosen patients for the trial instead of allowing all eligible patients to participate. And because these patients were seemingly selected, the trial results might not be indicative of what the results would be for other patients with the same lymphoma.

"But those patients we enrolled had such aggressive diseases, I'm pretty sure they wouldn't have been eligible for any other trial," Neelapu said.

It took a while for the skeptics to come around.

In 2017, data were presented with at least one year of follow-up. This time Neelapu presented PFS data for these patients. About half of the patients treated had a complete response and were disease-free for about fifteen months on average. Without CAR–T, only about 7 percent would have achieved a complete response.

There was little doubt anymore. CAR–T had shown its power in improving disease-free living for DLBCL patients with no other option.

Chapter 16

"You Are Cured"

FIRST APPROVAL

August 30, 2017
Basel, Switzerland

From their global headquarters, Novartis announced the first-ever CAR–T approval by the FDA. It was for their product, Tisa-cel, now named Kymriah, with which Emily was treated. The approval came five years after their partnership with the University of Pennsylvania, where Carl June had developed the CD19 CAR–T.

What started off with a little girl at her peril and then a miraculous recovery ended up showing a never-before-seen overall response rate of 83 percent in sixty-three patients with ALL whose disease had progressed after prior treatment. This FDA decision opened the door for a new approach to the treatment of cancer. The Oncologic Drugs Advisory Committee of the FDA unanimously recommended approval of Kymriah.

We fully expected this.

The FDA also approved tocilizumab to control the CRS that complicated the success of CAR–T. Before CAR–T, the patients in the Kymriah trial, both children and young adults, had received a myriad of chemotherapy, surgery, radiation, and targeted treatments or stem-cell transplant, and yet their five-year survival chances would've only been about 10 percent.

With CAR–T, we'd taken control of relapsing ALL brilliantly. The five-year, progression-free survival was about 50 percent in the Kymriah-treated patients. This was a dramatic increase.

Francis Collins, joint leader of the Human Genome Project and director of the NIH celebrated the occasion.

"Emily keeps smiling. Her smile gives me hope. Seeing her grow from a young girl struggling with a frightening disease into a poised woman who looks forward to starting 7th grade, but still spends time being an ambassador for immunotherapy, is one of the greatest joys I've had as NIH Director." Collins wrote in his blog on the same day Kymriah was approved.

Emily had been cancer-free for five years. Could we declare victory against ALL? This question was on my mind. In the cancer world, cure is referred to as the *C-word*. A medical team is always hesitant before declaring someone cured of cancer. But in Emily's case and for other children age five to nine, five years of complete remission verified CAR–T's ability to keep the body free from cancer for a long period.

We were almost there but needed to wait longer for the supreme outcome— the C-word.

Just about one month after Kymriah was approved, axi-cel, now called Yescarta, of Kite pharma received its approval for patients with large B-cell lymphoma who had failed two or more previous treatments. Overall response rate was more than 80 percent with Yescarta. The patients began to respond within thirty days of infusion and at a median, their response lasted for about eleven months. The PFS approached six months for these patients Neelapu and colleagues reported in the *New England Journal of Medicine*.

This was the beginning. In the next five years, we saw six more approvals for CAR–T-cell therapies, all in blood cancers. Of the six, four approved products were CD19–targeting and the other two CARs targeted a different protein on the plasma cell surface, *B-cell maturation antigen* (BCMA). In all instances, the approvals came for patients who'd already received and relapsed after prior treatment.

Just like other targeted treatment approaches, a key question I have is: Should we give CAR–T to patients who are newly diagnosed, who are yet to receive any kind of treatment for their cancer? Why wait if we could possibly cure a tumor with these cell treatments?

A COMPLICATED CHOICE

Administering CAR–T as a preliminary treatment in aggressive blood cancers is a tantalizing proposition.

We've seen incredible response rates from CAR–T, mainly in patients who didn't have many choices left. They've already received standard treatments such as combination chemotherapy, which produced remissions for a period. On some occasions, chemotherapy in combination with targeted drugs cures cancers. For example, when adults are diagnosed with ALL for the first time, chemotherapy can lead to complete remission in about 80 to 90 percent of the patients. In about half of these adults, chemotherapy cures the disease. That is a high rate of response, especially when one takes into consideration the availability, simplicity, and relative affordability of these conventional drugs.

On the other hand, if someone is diagnosed with an aggressive blood cancer and they've seen CAR–T doing amazing things, it is likely that they would want a treatment that, despite its complexity, will deliver a cure from the disease. Get a treatment and get cured—that's the simplest solution.

Except it isn't a simple decision.

First, CAR–T is not a plug-and-play treatment. A patient can't just get treated with CAR–T without detailed assessments, and there are major limitations in the process. This treatment involves highly complex technology to manufacture the cells. T-cells can only be harvested, transferred, genetically modified, grown, and infused by specialized centers with expertise in cell therapy. In the United States there are about one hundred CAR–T centers. This isn't enough. This immediately excludes a vast majority of our patients who either live far from a multidisciplinary CAR–T center or aren't able to get to a center to receive CAR–T. If we look beyond the West, there are thousands of patients in different parts of the world who aren't able to receive CAR–T because of technical and financial reasons.

Second, not everyone's eligible to receive CAR–T. Typically about three or more weeks are needed to manufacture the cells during which time the patient's cancer won't pause. This is especially true for newly diagnosed patients, who could greatly benefit from upfront chemotherapy with or without another targeted treatment to control their exploding cancer. They might not be able to wait while their CAR–T products are being produced.

If the patient is frail from other comorbidities or from an advanced cancer, they may not be able to weather a severe CRS or neurotoxicity with CAR–T. Just this fitness prospect alone would exclude a big chunk of our patients from the possibility of getting CAR–T. It is difficult to choose a first-line treatment that is costly and tough to produce; causes serious, life-threatening side effects; and involves a long waiting period when there are immediately available drugs or treatment regimens that are known to give long-lasting remission.

Then there's a practical question. If someone is treated with CAR–T as a first-line treatment and they don't achieve complete remission of their cancer (not all patients respond to CAR–T), what other second- or third-line treatment will be effective? Will the CAR–T be less effective later on given we have already used it up-front and caused the tumors to adopt evasion strategies?

A lot of these barriers might seem too difficult to overcome, but CAR–T superstar Neelapu begs to defer.

"Chemotherapy for lymphoma lasts for six months," he argues. "For leukemia, it can take more than a year. With CAR–T, it's just one infusion. A patient can go back to their usual activities, their life, sooner—isn't that a huge advantage?"

But Neelapu also agrees that the data are not enough currently to say CAR–T treatment is the best first-line choice. "It may be the case soon, though," he says. He is optimistic.

To help answer this, Kite Pharma has completed enrollment of forty patients with high-risk large B-cell lymphoma in their ZUMA-12 trial and treated these patients with CD19 CAR–T as a first-line treatment. Neelapu reports nearly 80 percent of the patients achieved a complete response in this trial. Although there was no head-to-head comparison of this treatment with another first-line treatment, similar lymphoma patients historically would only achieve about 50 to 60 percent complete response with chemotherapy.

CAR–T appears to be doing better than first-line chemotherapy in this small group of forty patients with aggressive lymphoma but this is not necessarily clear evidence of superiority. We need more investigations like this to find out the benefit of up-front CAR–T compared to standard, chemo-based regimens. My guess is that the standard treatment for now is the more logical first-line option for a majority of our patients.

That said, I'd love to see more of these bold approaches that challenge the status quo. CAR–T therapy may not be a reasonable option for every patient early in their disease, but it's becoming a strong choice as more data emerge.

"COVER ME IN SUNSHINE"

May 22, 2022

A bright seventeen-year-old Emily Whitehead stepped into the spotlight on the *Today Show* along with her parents and her oncologist Stephan Grupp.

"I am feeling great, I am feeling healthy!" she said with a smile. "I received my driver's license in January!"

"Something else to worry about," muttered her smiling father, Tom, sitting beside her.

Ten years after her CAR–T treatment, Emily has been cured of ALL. She is looking forward to taking the SAT and going to college.

Chapter 17

Escape from T

Promise me absolutely that you will leave, as fast as you can, if you begin to feel the house catching at you.

—Shirley Jackson

It is easy to be dazzled by the amazing responses seen with CAR–T-cell therapy. After decades of observing incremental improvements with anti-cancer drugs, it must seem to us that immunotherapy, including CAR–T, is a tremendous success. It changed cancer research in a huge way.

Is cell therapy the super weapon that's been missing from our arsenal? Is it the ultimate answer to cancer? It is often said that while immunotherapy, in general, is a broad category, CAR–T-cell therapy is a restricted option. Do we have enough evidence to claim this therapy is essential?

Genetically modified T-cells have been in the human system (those who received it) for a little over a decade now. In this time, we've been learning how these CAR–T-cells expand inside humans, how they affect their target cells, and how they get exhausted in the process. I have so far shared the glory of engineered T-cells against human blood cancers. I have not done justice to the whole story by laying out the uncertainty of having to deal with inadequate T-cell action and the possible dangers of tweaking the T-cell genome

So now, I want to begin that with William's story, which I believe says a lot about the kind of challenges patients and their families deal with.

ACHY FINGERS

June 2018
Washington, DC

When you are twenty-one, one of the last things that possibly comes to your mind is death. You like walking hand in hand with someone you desire, marveling at the monuments in our nation's capital, perhaps strolling in the city or lying down on the green, going to concerts, dining out, going to see movies—just enjoying life. William was in the city for an internship, and a new life awaited him.

Yet some achy fingers started bothering him. He couldn't have arthritis at this young age, could he?

It soon became much worse.

His throat became swollen and started to hurt badly. It felt like a terrible case of strep throat, but a bacterial infection was not supposed to be this acute. William was so exhausted that he would wake up for work and be tired enough to go back to bed by 10 a.m. The swollen throat took him to the emergency room. They diagnosed a viral infection and gave him a mouth-rinse that has a combination of several medicines to help improve the pain from his sore throat and mouth.

Back in his apartment, William stayed in bed for a few days. At one point he got up to grab a bottle of water in the kitchen. "My roommate was talking to me. I couldn't hear a word he was saying. I was in a delirium. I didn't know where I was really, and I didn't even have the strength to open our refrigerator," he said.

"Okay. We are going to the ER right now!"

The roommate acted promptly.

At the hospital they ran blood and urine tests. It was the blood-cell counts that gave William's cancer away. The doc came to his bedside with the results of the tests.

"You've been diagnosed with leukemia."

William didn't believe him. How could this be true? He took great care of himself, always prioritizing health over other things. No, it couldn't be true—he was in complete disbelief.

"Mom, I think the doctors are joking with me!" William called his mom and handed the phone to the nurse.

Disbelief from the shock of the diagnosis soon met with reality. Bedridden for a month, William was terrified. But he wanted to get better. Right after the diagnosis, he said to the medical team, "Let's just get this shit going." They ended up displaying this quote on his board at the hospital.

From the ER in Washington, William was transferred to a hospital in Baltimore. Because William's blood was swarming with leukemia cells, the oncologist decided on a round of chemo to fight back. William's family wanted him close to home in Indiana. As soon as William could walk, they transferred him to a regional hospital to continue his chemo.

To monitor his progress under the treatment, the oncologists did some scans and a bone marrow biopsy. This is a dreaded moment: you just don't know what the biopsy and the scans will reveal. It creates a lot of anxiety that can be tough to deal with.

For William, though, it was okay. "If you just say out loud the best-and-worst-case scenarios, it helps you ground yourself and realize where you are"—a bold coping strategy from a twenty-one-year-old.

In November 2018 with Thanksgiving around the corner, William completed his chemotherapy. But his bone marrow biopsy revealed leukemia cells in the marrow. His medical team had to find a different strategy before he got too sick again.

In December William had to make a decision quickly because his life possibly hung in the balance. He could go with a stem-cell transplant, which would be the conventional approach. Or he could go with a new cell treatment that targeted the CD19 present on the surface of B-type leukemia cells.

Remember, CD19 CAR–T for ALL had been approved at this point. But a lot more work needed to be done before a personalized CAR–T could be produced for William. Not all would agree this was the best course of treatment for him.

"I liked the idea of genetically modifying the cells," said William. "I liked being a small part of cancer research in a way. The main thing for me was you can do CAR–T and then bone marrow transplant but you can not do it the other way around."

This is strong reasoning again from a young man at a time when many of us would be overwhelmed by the sheer burden of the diagnosis. Another key factor for William in choosing the CD19 CAR–T over stem-cell transplant was the toxicity of the chemo. Before he could receive a bone marrow transplant he would have to get intense chemo and radiation as a preparation. He did not want the toxicity of these treatments again.

William made up his mind after talking with his medical team. They decided on CD19 CAR–T as his treatment.

While the T-cells were being prepared in a laboratory, back in the body from which they'd been obtained, the cancer cells kept on marching.

William's leukemia had to be kept to a manageable level until the CAR–T-cells were ready to be infused into his body. In mid-December William's CAR–T-cells were back from the lab. Typically, patients receive rounds of chemo for a few days and then a day of rest before receiving CAR–T-cells.

On day 5 William received his first CD19 CAR–T infusion.

The CAR–Ts created from his own cells were stored in small bags cooled by liquid nitrogen. Before infusion the bags were heated in a fluid bath then hooked up to the chemo-port attached to William's body. The infusion takes place by force of gravity so the drip happens by itself.

"It's very anticlimactic for something that you're hoping will save your life, honestly. It was a really cool experience nonetheless," William recalled.

I've heard patients talk about their calmness and a vitalizing sense when they received CAR–T. I've heard expressions such as "It's like tiny soldiers entering my body to fight the cancer cells." I love the tiny soldiers imagery: millions of living cells reentering the battleground after training with the purpose of defeating the enemy.

A month after William's CAR–T infusion, his medical team decided to check his bone marrow. On January 10, the biopsy results came back and William was in complete remission of acute leukemia. There were no leukemia cells in his marrow—a single infusion of CD19 CAR–T had effectively wiped out his leukemia.

This was consistent with other CAR–T stories. What came after was not. Six months later, William's leukemia came back.

It was jarring, it was heart-breaking. Surely after complete remission with CAR–T anyone in the young man's situation would have been certain they had defeated leukemia. After all, that's what we seemed to hear from the news stories.

"I checked all the boxes for staying healthy. I thought cancer was a chapter of my life I could close. I had a great six months of life without it, but it returned nonetheless."

I could hear the emotion in his statement. It is a devastating thing to deal with a cancer diagnosis for the first time. It's perhaps even more crushing to know that your cancer has returned after one of the most wondrous treatments that science can offer had apparently brought it to a complete remission.

Why did William's acute leukemia come back? What could be said about his CAR–T-cells that failed to keep up the fight? Could we have predicted this?

IMMUNE ESCAPE

The failure of CAR–T could be attributed to an immune-escape phenomenon. Tumor cells interact with their microenvironment, the tumor's milieu. In this environment, after an infusion of CAR–T, the tumor cells suddenly find themselves overwhelmed by an invading army of invigorated T-cells—a form of aggression they don't normally deal with. The genetically modified CAR–T-cells seek out their target cells very efficiently. If we take CD19–hitting CAR–T (this is the dominant CAR in the market) as an example, suddenly this CD19 becomes an existential threat to the tumor cells. What do they do to avoid being killed?

They alter their identity.

From our experience using CD38–targeting monoclonal antibodies, which I highlighted before, high levels of CD38 expression in myeloma cells ensures that these cells get killed first. But soon the expression of CD38 becomes low in the myeloma cells as a response to being hit by the antibody. Something similar can be seen in the blood of CLL patients treated with another antibody called rituximab. This antibody targets the CD20 protein on the surface of lymphocytes, and as a response to the drug, the cells start to diminish and even stop expressing CD20 temporarily.

So diminishing the expression of a target antigen is a known escape route for tumor cells. Having selectively been pressured by the CAR–T, the cancer cells do this ultimate adaptation—they lose the expression of the target antigen altogether.

Neelapu and his colleagues collected samples from patients in the ZUMA-1 trial who first responded to CD19–targeting CAR–T and who later relapsed like William after achieving remission. They checked for B-cell identity—CD19, CD20, and a few other proteins—in samples that had been collected before treatment and after relapse.

In one-third of the patient samples after cancer relapse, expression of CD19 had been lost. Interestingly, despite the loss of CD19 expression, all other B-cell proteins tested for, including CD20, were preserved. Given this, it is easier to understand why CD19 CAR–T would fail to keep the cancer at bay in these cases: its primary target was gone. The ever-elusive immune system had to play its tricks.

I don't know if CD19–expression loss was the reason behind the relapse from CAR–T therapy for William that first time, but one can't help wondering if it might have played a role. What would his medical team do at this point?

They went in a somewhat unprecedented direction.

Six weeks after his relapse, William received another round of the same CAR–T infusion. This time there was more anxiety, more speculation. Would the leukemia come back a third time? And if it did, what would be the course of action?

This was difficult to say.

William was one of very few (only the second patient at Johns Hopkins) to have received CAR–T twice. I was intrigued by the idea that a second CAR–T infusion could completely work, but I couldn't help wondering what would be so different about the second infusion. This was the very same CD19 CAR he'd received before. Would it work now?

Well, the answer may lie in the context of tired T-cells.

Back to the tumor milieu then.

CAR–T cells have just been infused. Now we have the CAR–T-cells engaged with the tumor cells, invading their space, and killing them in a hit-and-run fashion: enter the space, eliminate the target, leave the scene.

Then repeat.

At this rate of work, the CAR–T-cells can become less efficient. They are said to become exhausted from all this action after a period. How do we know if the T-cells are tired?

There are inhibitory proteins that if present signal a reduced responsiveness of the CAR–T-cells. So if these markers, PD1, LAG3, TIM3, are expressed, this indicates tired T-cells.

If an army is fatigued from a fierce battle, what's a good remedy? Reinforcement. Fresh boots on the ground. That's one idea suggesting a second infusion of CAR–T-cells might be successful. With a new injection of new CAR–T, perhaps the tired T-cells will have the necessary ammunition to finish the job.

There is another way for tumor cells to escape T-cells. In blood cancer, bone marrow plays a critical role. This is where blood cells are produced, and this is the place that gets taken over by the tumor cells so brazenly. Predictably, the bone marrow tumor milieu communicates with the malignant cells. As a result of this communication, tumor cells bolster their *anti-apoptotic* work-stream so that they don't initiate a programmed demise (*apoptosis* means cell death that takes place as a part of a cell's normal development). There are other bystander cells in the bone marrow tumor milieu that can help suppress the action of CAR–T-cells. Our understanding on this topic is incomplete.

William wasn't afraid. He was playing the hand he'd been dealt. The second CAR–T infusion gave him a complete remission.

Chapter 18

CAR and Beyond

With unprecedented results comes unprecedented cost. Ever since the first approval of CAR–T-cell therapy in 2017, the high cost of this therapy has been a major barrier.

CAR products can be priced between $373,000 to $475,000 per infusion just to acquire the cells. There are additional costs for using a facility and other procedures. Hospitalization and office visits can cost upward of $53,000 in academic centers specialized in oncology. For nonacademic settings, this hospital visit cost can be less—around $23,000.

Then there is a high chance of CRS for these patients. The management of CRS can range from $30,000 to $56,000 per patient. The grand total for CAR–T therapy could, therefore, be a staggering $500,000! I've also seen estimates of CAR–T costing more than $1 million for the total cost of treatment.

Is the high cost worth it?

That is a difficult question to answer. We need more variables, better comparative assessments that can support a cost-effectiveness analysis.

Do the better response rates and longer survival outweigh the astronomical costs of CAR–T therapy? As I mentioned before, these CAR–T trials were mainly done without a head-to-head comparison with other treatments. Typically an experimental treatment is compared with the standard treatment in a *randomized controlled trial* (a direct comparison), but in the case of CAR–T, this is not always feasible. Thus any comparison with other treatments could be indirect.

John Lin and his colleagues undertook a cost-effectiveness analysis of the first approved CAR–T product, tisa-cel, for children with ALL. Tisa-cel had the longest follow-up period among all the CAR–T products, a median of about forty months, and so is a good example for such an analysis.

What they found might make you think harder.

Lin et al. assumed tisa-cel would lead to relapse-free living of at least five years for 40 percent of all children who were treated with it. Yes, this was an optimistic assumption, but let's just go with it for now. If that's the benefit, tisa-cel would lead to a survival gain of twelve years over other non-cell treatments. For each year of a child's life, the cost gain was estimated to be about $61,000. So if a child survived for twelve years with CAR–T, the total saved would be about $730,000.

Recall from our earlier estimate that a one-time CAR–T infusion can cost about $500,000. So we can say in this case that if CAR–T really delivers a five-year relapse-free survival, then getting it is worth it despite the high cost of the product.

However, the devil is in the details. Five-year relapse-free survival isn't a certainty. The reality is it might be achieved in only a few children. If only 20 percent (or fewer) of the children who receive this CAR–T live relapse-free for five years, the value of the treatment decreases and the cost of getting it is difficult to justify from an insurer's perspective.

The bottom line is that if the cost to get CAR–T can be reduced, it can certainly become more attractive and sustainable even if we assume less-than-stellar clinical benefits. The current model of personalized CAR–T may not see a dramatic decrease in cost anytime soon. But there are other, newer versions of cell therapy that could offer relief from a cost standpoint.

So far the CAR–T-cell treatments that have been successful in blood cancers have all used *autologous* T-cells, meaning the T-cells were collected from the patient. After harvesting the T-cells from the patient, their CAR–T is genetically engineered. But if someone doesn't have enough T-cells for CAR engineering, they are out of luck.

Well, not quite.

Just as in stem-cell transplant, autologous is a great approach but there is another choice: *allogeneic*. Here *allo* means "other." So instead of T-cells gleaned from the patient, the T-cells can be collected from someone other than the patient.

In this approach, the T-cells are harvested from healthy donors—an advantage because a healthy giver is more likely to have better-performing T-cells as these cells have not been affected by cancer or drugs like chemotherapy. The main advantages of using donor T-cells, though, are availability, time, and cost. From a single donor, multiple batches of CAR–T products can be produced quickly giving patients an immediate chance to receive treatment. There's no need to wait for weeks to manufacture an individualized autologous CAR–T. The cost for allogeneic CAR–T can be significantly less because these cells can be produced on a larger scale.

This would be an off-the-shelf version of CAR–T.

But there's a serious concern with these allogeneic CAR–T-cells and this is tied to the topic of graft rejection—an immune phenomenon we discussed earlier. In this off-the-shelf CAR–T, the T-cells are from a donor and therefore are seen as non-self by the immune system in the patient's body. The system wants to eliminate these non-self T-cells through hostile rejection, known as *graft-versus-host disease*. It's a life-threatening condition and could very well outweigh the anti-tumor activity of the allogeneic CAR–T. Scientists are working to find ways to overcome the rejection challenge of donor CAR–T.

In one such approach, researchers in China designed donor-derived CAR–T-cells with a genetic modification to limit immune rejection of allogeneic T-cells. They treated twelve patients in a Phase 1 trial with this allogeneic CAR–T and no one suffered from graft rejection or serious CRS or neurotoxicity. More than 60 percent of the patients showed a complete response to their cancer—leukemia and lymphoma—at the end of one month after treatment, boosting confidence in the possible usefulness of off-the-shelf versions of CAR–T therapy. Multiple international pharma companies are now racing to conduct similar trials and bring their off-the-shelf CAR–T products to the market. About 125 allogeneic CAR–T cells are in clinical trials now.

Our patients need curative, novel treatments that are patient-friendly, and we have a strong option in the works.

BISPECIFIC BANDWAGON

Among my first assignments in the drug-making industry (alongside mAb85), I took an asset that wasn't a small molecule, monoclonal antibody, or a CAR–T to its first-time-in-human trial. It was a new class of drugs known as *bispecific T-cell engagers* (TCE; in reality this class of drugs could also be trispecific, meaning the molecule can bind to two, three, or more targets).

As our quest against blood cancer saw a slew of CAR–T approvals, TCEs followed in the same footsteps in immunotherapy, using the power of the immune system to attack tumor cells, except that instead of using engineered T-cells, this approach recruits T-cells, patrolling in the body without harvesting or altering them. Unlike CAR–T, TCEs are more conventional drugs. They activate the patient's own T-cells and redirect these killer cells inside the body while holding on to the cancer cells. Since I mentioned "holding on to," let's clarify this a bit.

TCEs are modified mAbs. These antibodies, we know, are *Y*-shaped structures. A naturally occurring antibody usually is specific for one antigen—a specific site located at the tip of each arm of the antibody—and attaches to that antigen. But antibodies are also very moldable. Instead of being distinct for one antigen, a bispecific therapeutic antibody can be specific to two or more different antigens simultaneously.

Consider the shape of a *Y*. One arm of the *Y* attaches to an antigen while the other arm can be engineered to detect one or more other antigens, simultaneously holding on to different proteins on the surface of tumor cells as well as immune cells like T-cells. If the tip of one arm of an engineered antibody attaches to an antigen on the surface of a cancer cell, the other arm can recruit T-cells in the vicinity and also boost these recruited T-cells via a gas pedal like CD28. In this way not only can a TCE crack down on cancer cells via direct inhibition, it can stimulate nearby T-cells to join in the fight.

This is just one broad example of how a TCE can be designed. There are hundreds of variations on its scheme but the goal is the same: harness the power of T-cells (and other immune cells) without the need to extract, genetically engineer, and reinfuse T-cells with the same end result, that is, killing off the enemy, the tumor cells, efficiently.

When I kicked off developing a TCE in 2018, my background research in the space yielded two things: first, TCEs had been a long time emerging, and second, almost every drugmaker I knew was developing one or more of these TCEs! In 2018, the majority of these TCEs were in early clinical trials with just one FDA–approved drug.

Right now, we have at least seven drugs approved in this class, and six of them have been approved since 2019. That's almost two approvals per year heating up the race for becoming the best-in-class medicine. About two hundred clinical trials with TCEs are ongoing now for a variety of cancer subtypes, and perhaps one hundred more about to begin a trial.

But what does this mean for our patients?

These TCEs are a great answer to the difficulties with CAR–T-cell therapy. In contrast to autologous CAR–T, TCEs are off-the-shelf, universal, easily given, and have better side-effect profiles than CAR–T. From a cost-effectiveness perspective, there is no question that TCEs are more affordable. These drugs cost much less: about $5,000 for one injection with a total cost of treatment approaching $65,000 (this is an estimate). Unlike CAR–T, however, TCEs need to be given in multiple doses, and the full treatment can be several months long.

Will these give long remissions from blood cancers and provide long-term, disease-free living for our patients? Can they match the amazing response rates and even cures of CAR–T? How much better or worse will TCEs be if given in combination with standard treatments?

Early data suggests a substantial improvement in disease remissions for several blood cancers like leukemia, lymphoma, and myeloma with these TCEs. The next few years will answer questions of their long-term survival benefits and dictate their legacy.

Will TCE be a clearly better treatment choice than CAR–T?

While the answer to this question is still pending, I'm looking forward to continuing this path of developing T-cell engagers in the blood cancer space, and I am confident of great benefits from this novel class of drugs.

Chapter 19

What's Next?

Whether we will acquire the understanding and wisdom necessary to come to grips with the scientific revelations of the twentieth century will be the most profound challenge of the twenty-first.

—Carl Sagan

So here we are in the twenty-first century. At the dawn of this millennium, we began uncovering the mysteries of our own DNA, and while appreciating the similarities between ourselves no matter how different we may look, we've realized how seemingly trivial differences in our genes can lead to diseases with profound consequences.

As we advance into the twenty-first century, reducing the burden of many infectious diseases (and eliminating at least two) thus far in our journey, even so, one out of every two males and one out of every three females will develop one type of cancer or another in their lifetime. About one in five will die of their cancer. The numbers are from lifetime risk of cancer as calculated by the American Cancer Society for the U.S. population based on recent cancer data. The lifetime cancer risk is similar in other industrialized nations.

In this century, cancer will be the number one cause of death worldwide. We are getting older and surviving longer. Cancer is a disease of errors in the instructions in our genes, and over time these errors build up. The longer we live, the more likely it is for the faults in our DNA to show up as cancers.

Yes, we've improved survival from cancer. In fact, we've doubled survival rates in the past fifty years but the reality of it is that we won't be able to keep treating this rising population of newly diagnosed patients who'll have various types of cancers. No treatment discovery by itself will be able to smother the ballooning burden of cancer. This is a disease that leaves an ineradicable trail of despair.

We need to do a better job at prevention.

We need to firmly clamp down on the known, avoidable causes of cancers. Smoking, alcohol, and obesity—we know these have a direct relationship with the king of terror, yet over the past decade, somehow, the most important message that prevention is the best medicine hasn't gotten through. Preventive measures aren't cheap, but they're a better option than trying to come up with innovative but costly treatments. The other form of prevention, early detection, needs to be done more thoroughly.

A fifty-year-old may have had their first cancer cell form months ago, undetected, before they finally show up with a fully metastatic disease. The time between first appearance of cancer cells and end-stage disease is what our health system needs to shorten to meaningfully improve cure rates. Of course the sooner a diagnosis is reached, the earlier intervention can be begin. For early forms of blood cancer, a shorter lag time between diagnosis and the start of immunotherapy translates into a decreased chance of the cancer's progression.

I anticipate technological advances in early detection of multiple cancers relatively soon. These detections will depend on *biomarkers*: proteins, DNA, RNA, cell-free DNA, T-cells, and other minute bits of evidence of cancer in its early form. *Minimal residual disease* (MRD) detection methods will apply here, and their ability to detect cancer when it's present (i.e., their sensitivity and predictability) will improve. Currently these tests can fish out one tumor cell from a crowd of 1 million normal cells, and the sensitivity is expected to be even better soon.

Blood-based genomic technologies are swiftly transitioning into a multi-cancer detecting platform that will possibly identify at the same time the risks for several cancers. This poses a quandary on how to interpret such results within the context of conventional diagnostic approaches. I am aware of at least five newer early cancer detection tests that promise to uncover traces of multiple cancers, all of them collected from blood early in the disease and potentially offering several years of lead time. I'm sure there will be more such methods and technologies available soon. We need to use these tests and integrate their findings with clinical trials and more medical records data.

While early detection of budding cancer is great, it does pose the problem of overdiagnosis. If our tests are too sensitive, we'll pick up too many supposed cancers, which would be false alarms. As a direct consequence of overdiagnosis, some medical practitioners will take it one step further and consider treating based upon such results.

Very recently I was involved in an interesting discussion on treating a precursor form of multiple myeloma, known as smoldering myeloma. The argument here was to treat smoldering myeloma before it could develop into full-blown myeloma so that the patient wouldn't need the standard treatment. The premise was to treat this patient with CAR–T or conventional chemotherapy. This raised a few eyebrows. CAR–T therapy isn't an automatic choice even in newly diagnosed, aggressive cancers, but it was being offered here for a precursor condition. I understand the intention: eradicate a possible cancer early before it matures, but one can't help wondering if CAR–T might be too bold an approach here. Nonetheless this is a path that is attractive to a lot of minds: how best to intervene in precursor cancers.

What about treatment? How will new cases be treated with novel drugs and cell therapies? In the first decade of the twenty-first century, we were busy adding one treatment to another, often using a known drug with a new drug and comparing the results versus results with the standard treatment alone. This additive model of old versus new + old starts to lose its appeal when you look at all the new drugs that are becoming available. It will require a huge number of patients and many iterations of trials to prove the benefit of a new drug if we stick with conventional clinical trial designs.

In the next few years, I'm expecting an optimized way to test the potential benefits of new medicines wherein a patient's cancer biology will dictate which drugs to use and when. More new + new combinations of therapies will be suggested, and I anticipate regulators will be open to this. For example, bispecific antibodies will push the frontier in combination with other bispecifics or with other small molecules.

Cell therapies like CAR–T will continue to dominate the immunotherapy space. CAR–T-cells will be given advanced functionality as we continue to genetically engineer cells. So far CAR–T has not been successful with solid tumors but that may start to change with refinement of cells via gene editing. Off-the-shelf CAR–T data will be released in the next several years and it could be a game-changer.

I expect to see a better, more-seamless integration of our multi-omics platforms—for example, genomics, proteomics, metabolomics—with the discovery and development of drugs. As vast data sets become available on the multi-omics, artificial intelligence (AI) will play a critical role in picking up patterns in the data, correlating between different -omics, and predicting which drug targets will offer the best efficacy in a given cancer. AI–driven predictive models for a drug's ideal dose, safety, and benefits could very well become a supporting feature of every new drug's investigation and licensing application in the future.

Sometimes I can't help wondering whether we could, with our technological ingenuity, emancipate ourselves from clinical judgment and adopt a machine-taught, non-intuitive algorithm to tackle cancer. I don't know if this is possible, but I'm happy to be a drug developer at a time when we're witnessing a technological revolution. We will cure a handful of standard-risk cancers through a combined human-machine endeavor, and that day could come sooner than we think.

Epilogue

On a personal level, I have now looked cancer in the face at least twice. The king of terror has seemingly averted its stare, allowing me to pass. Something tells me I am not done with this matchup. But I've chosen my path: I will engage with some of the brightest minds in science and clinics every day to find new medicines and therapies for cancer.

There are scientific questions that I am eager to know the answers to, such as in what specific ways the tumor milieu changes its behavior to allow a cancer cell to grow. I want to learn what signals we can pick up early from this microscopic environment to anticipate cancerous formation. And how quickly will the current path of drug discovery and development adjust to the wealth of data we are generating and translate to new therapies for cancer?

I am but an infinitesimal cog in this vast, global, scientific workshop that has the capacity and the talent to answer these questions. We are standing upon a very strong foundation built by the pioneers of science and medicine, some of whom are featured in this book.

As for our modern-day protagonists, Craig Venter has found his next target: defying death. Moving on from his short stint in drug development, Venter cofounded Human Longevity, a venture that aims to integrate knowledge of genomics (and other -omics), personalized diagnostic tests, precision medicine, and artificial intelligence (AI) to lengthen life. The initiative has a program called 100+ meant to extend human life beyond one hundred years. At the center of it is a scientific paper published in 2020 in which the group determines that about one in six adults have at least one genetic anomaly. When these genetic variants are correlated with observable findings, about 30 percent to 100 percent adults have strong associations between them—the genotype and the phenotype.

"For individuals with life-altering findings, the projected additional life extension year is 8.51 years on average," the group claims. So via thorough physical examination, diagnostic imaging, ultrasound, cognitive tests, biological samples testing, and DNA analysis, the Human Longevity Initiative's proprietary program has claimed to have actually increased lifespan for some individuals! A very Venter-isque, grand vision to deliver on the human genome's promise.

Richard Miller cofounded Corvus Pharmaceuticals and currently serves as the CEO of the company based in San Mateo, California. As an adjunct professor, Miller still spends a day each week seeing patients in the Lymphoma Clinic at Stanford University. He won the Drug Hunter Award in 2020. At Corvus, Miller is championing small molecule inhibitors as well as antibodies. Soon after the Drug Hunter Award announcement, Miller said, "A lot of people can look good for one day, but you want to sustain excellence over an extended period of time." That is a vintage Miller statement: clever, hinting at more than one thing at a time. Of course, his achievements will continue to impress us.

Like Miller, Ahmed Hamdy cofounded a company, Acerta Pharma, which later became part of AstraZeneca. After being booted from Pharmacyclics, Hamdy played a pivotal role in getting acalabrutinib (another BTK inhibitor) approved. He now serves as the CEO and chairman of Vincerx Pharma with his long-time partner Raquel Izumi as COO.

True to his commitments, Jeff Sharman continues to practice in the heartland of America. Every morning, Sharman starts his day with a drive from his home to a community oncology center in the beautiful Willamette River Valley of Eugene, Oregon, to take care of his patients. He brings great perspectives from his community and forms an important bridge with academic centers and drug developers like me. Sharman's expertise and insights on not just blood cancers but pretty much all types of cancers is valued widely. Last I checked, he was contemplating writing a book (although after hearing me whine about meeting the deadline, he seemed somewhat discouraged). I sure hope he changes his mind!

Michael Wang is an endowed professor with the Department of Lymphoma and Myeloma at the MD Anderson Cancer Center in Houston, Texas. After playing critical roles in the approvals of ibrutinib and acalabrutinib, Wang led the investigation of the first CAR–T therapy in mantle cell lymphoma. In 2020 this product was approved by the FDA. What's more, Wang played a key role in the approval of yet another BTK inhibitor drug for MCL, pirtobrutinib, in 2023. Save for possibly Ogden Bruton himself, Wang has become synonymous with BTK. With renewed vigor, he shows up at work every day with the same infectious smile on his face.

Several doors down the corridor from Wang's office on the sixth floor of the faculty center at MD Anderson Cancer Center sits Sattva Neelapu, professor and deputy chair of the department. I had the audacity to ask him if he was interested in taking part in a new trispecific T-cell engager drug (asking a cell therapy trailblazer for taking on a T-cell engager—cringeworthy, I know), to which Neelapu coldly replied, "I am only doing CAR–T now." In his laboratory, Neelapu is experimenting with the next generation of CAR–T: cell therapies that target more than one tumor antigen.

Simon Rule and I now share the same workplace and meet almost every week. Besides discussing strategies for practical drug development, we talk about the game of cricket a lot! Whenever I am with Rule, I stay alert because you never quite know when he's gonna say something quotable. I have a lot of respect for his no-nonsense wisdom.

George Klein passed away in 2016, at the age of ninety-one. Looking back on when George met Eva, I cannot help but wonder what would have happened had the young man not been forced to take a vacation on Lake Balaton and had he not met Eva. Would he have become an immunologist? Would the couple have pursued research the way they did?

One wonders.

Emily Whitehead began her first semester at the University of Pennsylvania as a freshman in November 2023, a campus she had visited eleven years ago to receive her groundbreaking CAR–T treatment for leukemia. Looking out the window of her dorm, Emily could see the hospital where she was treated. Hers is an incredibly inspirational story that keeps on getting better.

As I make changes in proof for this book (which has become a tedious process), my train jerks forward, leaving the main station in Frankfurt. After attending the European Hematological Association's annual meeting there, I'm going on to a bi-annual meeting in Lugano, Switzerland. The express train quickly gathers momentum, but the difference in transatlantic time zones, coupled with the stress of travel, gets the better of me. I sleep for a few hours, and when I finally wake up, I can see through the window the station we're in: Freiburg—Paul Ehrlich's early workplace!

So I step off the train. Lugano can wait a few more hours.

Makhdum Ahmed
June 13, 2023
Freiburg, Germany

Acknowledgments

When I conceived of the idea of this book, I was working 110 percent—there was really no time to invest in another project. And then the COVID pandemic hit. We were in lockdown just like millions of others here in the United States and around the world, and it allowed me extra time in the evenings and late nights to focus on my writing. My wife and partner of fifteen years, Dilruba A. Zakia, has been a great source of support. So were my three children Maya, Myreen, and Manseeb. The difficult part for my children was that the pandemic eventually ended and the lockdown was lifted, yet their dad still appeared red-eyed and unmindful from his nighttime writing endeavors now coupled with his daytime job. This young family is a gem. I can't wait to watch my children grow and hopefully do amazing things for humanity.

Huge thanks to my agent Linda Konner, who read my first book proposal, tore it apart, and developed it into something acceptable. Without Linda, this book would not have been possible. Special thanks to Whitney Noziska for your careful comments, edits, and suggestions. My editor Jake Bonar's youthful vigor helped me meet the timeline (only barely).

To my colleagues, friends, mentors, and teachers, I am greatly indebted to you for taking the time to chat with me and help me craft this narrative. My thanks and gratitude to Michael Wang, Simon Rule, Jeff Sharman, Saad Usmani, Sattva Neelapu, Paul Richardson, Nina Shah, Prashant Kapoor, Tak Mak, Helgi van de Velde, Maria Badillo, Sarosh Effendi, and Jamal Saeh for your words of wisdom, suggestions, and expertise. Thanks to Lara Marks for your kind gesture for the monoclonal antibody section. I am a fan of Stephanie Chuang's works at https://thepatientstory.com. She is a brilliant advocate for our patients, and a few of the stories there are featured in this book with Stephanie's kind permission.

Acknowledgments

References

PROLOGUE

David Scadden. *Cancerland: A Medical Memoir*. New York: Thomas Dunne Books, 2018.

Peter Wehrwein. "Propofol: The Drug That Killed Michael Jackson." Harvard Health Publishing. www.health.harvard.edu/blog/propofol-the-drug-that-killed-michael -jackson-201111073772.

CHAPTER 1

The Wondrous Database

Andrew Pollack. "U.S. Hopes to Stem Rush toward Patenting of Genes." *New York Times*, 2000. https://archive.nytimes.com/www.nytimes.com/library/national /science/062800sci-genome-patents.html.

P. M. Rowe. "Patenting Genes: J. Craig Venter and the Human Genome Project." *Molecular Medicine Today* 1, no. 1 (1995): 12–14.

U.S. Department of Energy. "Human Genome Project Information Archive 1990– 2003." www.ornl.gov/hgmis.

Craig J. Venter. *A Life Decoded: My Genome, My Life*. New York: Penguin Books, 2007.

White House, Office of the Press Secretary. "The Human Genome Project: Benefiting All Humanity." https://clintonwhitehouse3.archives.gov/WH/New/html/20000315 _3.html.

H. L. Williams. "Intellectual Property Rights and Innovation: Evidence from the Human Genome." *Journal of Political Economy* 121, no. 1 (2010): 1–27.

The Workshop of a Chemotherapist

F. Bosch and L. Rosich. "The Contributions of Paul Ehrlich to Pharmacology: A Tribute on the Occasion of the Centenary of His Nobel Prize." *Pharmacology* 82, no. 3 (2008): 171–79.

George Dunea. "Paul Ehrlich: From Aniline Dyes to the Magic Bullet." Hektoen Institute of Medicine. https://hekint.org/2018/10/10/paul-ehrlich-from-aniline-dyes -to-the-magic-bullet/.

Paul Ehrlich. *The Collected Papers of Paul Ehrlich*. Edited by F. Himmelweit. Oxford: Pergamon Press, 1960.

V. C. Jordan. "50th Anniversary of the First Clinical Trial with ICI 46,474 (Tamoxifen): Then What Happened?" *Endocrine-Related Cancer* 28, no. 1 (2021): R11–R30.

A. B. Kay. "Paul Ehrlich and the Early History of Granulocytes." *Microbiology Spectrum* 4, no. 4 (2016).

Robert S. Schwartz. "Paul Ehrlich's Magic Bullets." *New England Journal of Medicine* 350, no. 11 (2004): 1079–80.

Michael Titford. "Paul Ehrlich: Histological Staining, Immunology, Chemotherapy." *Laboratory Medicine* 41, no. 8 (2010): 497–98.

P. Valent, B. Groner, and U. Schumacher, et al. "Paul Ehrlich (1854–1915) and His Contributions to the Foundation and Birth of Translational Medicine." *Journal of Innate Immunity* 8, no. 2 (2016): 111–20.

K. J. Williams. "The Introduction of 'Chemotherapy': Using Arsphenamine— The First Magic Bullet." *Journal of the Royal Society of Medicine* 102, no. 8 (2009): 343–48.

Yale Medicine Magazine. "From the Field of Battle, an Early Strike at Cancer." https: //medicine.yale.edu/news/yale-medicine-magazine/article/from-the-field-of-battle -an-early-strike/#:~:text=At%20the%20start%20of%20World,first%20effective %20chemotherapy%20for%20cancer.

Celera's Quest in Creating a New Drug

BioSpace. "Pharmacyclics, Inc. Buys Drug Candidates from Celera Genomics." www.biospace.com/article/releases/pharmacyclics-inc-buys-drug-candidates -from-celera-genomics-/#:~:text=Under%20the%20terms%20of%20the,kinases %20involved%20in%20immune%20function.

Carina Dennis. "Venter's Departure Sees Celera Seek Therapies." *Nature* 415, no. 6871 (2002): 461–61.

Denise Gellene. "Co-Founder Venter Steps Down as President of Celera Genomics." *Los Angeles Times*, 2002. www.latimes.com/archives/la-xpm-2002-jan-23-fi -venter23-story.html.

Scott Hensley. "Craig Venter Leaves Celera as Firm Seeks New Direction." *Wall Street Journal*, 2002.

Daniel J. Kevles and Ari Berkowitz. "The Gene Patenting Controversy: A Convergence of Law, Economic Interests, and Ethics." *Brooklyn Law Review* 67, no. 1 (2001): 233.

Pharma Letter. "Celera Genomics to Acquire Axys in $174 Million Deal," 2001. www .thepharmaletter.com/article/celera-genomics-to-acquire-axys-in-174-million-deal.

Andrew Pollack. "The Genome Is Mapped: Now He Wants Profit." *New York Times*, 2002. www.nytimes.com/2002/02/24/business/the-genome-is-mapped-now -he-wants-profit.html.

David Shaywitz. "The Wild Story Behind a Promising Experimental Cancer Drug." *Forbes*, 2013. www.forbes.com/sites/davidshaywitz/2013/04/05/the-wild-story -behind-a-promising-experimental-cancer-drug/?sh=95321aa5857c.

J. C. Venter, K. Remington, and J. F. Heidelberg, et al. "Environmental Genome Shotgun Sequencing of the Sargasso Sea." *Science* 304, no. 5667 (2004): 66–74.

CHAPTER 2

The Promise

Brian J. Druker. "The Immaculate Conception of Gleevec, as Told by Brian Druker." *Genetic Engineering and Biotechnology News*. www.genengnews.com/gen-edge /the-immaculate-conception-of-gleevec-as-told-by-brian-druker/.

The Other Molecule

Manfred Kraus, Marat B. Alimzhanov, Nikolaus Rajewsky, and Klaus Rajewsky. "Survival of Resting Mature B Lymphocytes Depends on BCR Signaling Via the Igα/B Heterodimer." *Cell* 117, no. 6 (2004): 787–800.

M. I. Merolle, M. Ahmed, K. Nomie, and M. L. Wang. "The B Cell Receptor Signaling Pathway in Mantle Cell Lymphoma." *Oncotarget* 9, no. 38 (2018): 25332–41.

Jeff Sharman. Author interview, 2023.

Proof of Concept

Jonathan W. Friedberg, Jeff Sharman, and John Sweetenham, et al. "Inhibition of Syk with Fostamatinib Disodium Has Significant Clinical Activity in Non-Hodgkin Lymphoma and Chronic Lymphocytic Leukemia." *Blood* 115, no. 13 (2010): 2578–85.

Richard Murphey. "Biotech from Bust to Boom." Bay Bridge Bio, www.baybridgebio .com/blog/biotech-bust-to-boom.html.

CHAPTER 3

Miller's Misadventure

BioSpace. "Pharmacyclics, Inc. Cancer Drug Not Approvable Says U.S. FDA," 2007. www.biospace.com/article/releases/pharmacyclics-inc-cancer-drug-not -approvable-says-us-fda-/.

M. P. Mehta, P. Rodrigus, and C. H. Terhaard, et al. "Survival and Neurologic Outcomes in a Randomized Trial of Motexafin Gadolinium and Whole-Brain Radiation Therapy in Brain Metastases." *Journal of Clinical Oncology* 21, no. 13 (2003): 2529–36.

Richard Miller. "Cancer Regression." *Wall Street Journal*, 2007.

The First Trial

Jack Davis. "Docu-Drama: Shake-up at Top of Pharmacyclics." *Mercury News*, 2008. www.mercurynews.com/2008/09/19/docu-drama-shake-up-at-top-of -pharmacyclics/.

Silicon Valley Business Journal. "Pharmacyclics CEO and 4 Directors Resign, CFO to Depart," 2008. www.bizjournals.com/sanjose/stories/2008/09/08/daily83.html.

From the Brink

Jan A. Burger, Susan O'Brien, and Nathan Fowler, et al. "The Bruton's Tyrosine Kinase Inhibitor, PCI-32765, Is Well Tolerated and Demonstrates Promising Clinical Activity in Chronic Lymphocytic Leukemia (CLL) and Small Lymphocytic Lymphoma (SLL): An Update on Ongoing Phase 1 Studies." *Blood* 116, no. 21 (2010): 57.

Susan M. O'Brien. "PCI-32765 Is One of the Most Exciting Drugs Around." Practice Update and *Oncology*, 2012. www.practiceupdate.com/content/pci-32765-is-one -of-the-most-exciting-drugs-around/17488.

Susan O'Brien, Jan A. Burger, and Kristie A. Blum, et al. "The Bruton's Tyrosine Kinase (BTK) Inhibitor PCI-32765 Induces Durable Responses in Relapsed or Refractory (R/R) Chronic Lymphocytic Leukemia/Small Lymphocytic Lymphoma (CLL/SLL): Follow-Up of a Phase IB/II Study." *Blood* 118, no. 21 (2011): 983–83.

Nathan Vardi. *For Blood and Money: Billionaires, Biotech, and the Quest for a Blockbuster Drug*. New York: W.W. Norton & Company, 2023.

Step on the Gas

S. Ponader, S. S. Chen, and J. J. Buggy, et al. "The Bruton Tyrosine Kinase Inhibitor PCI-32765 Thwarts Chronic Lymphocytic Leukemia Cell Survival and Tissue Homing in Vitro and in Vivo." *Blood* 119, no. 5 (2012): 1182–89.

U.S. Food & Drug Administration. "Accelerated Approval Program," 2023. www.fda
.gov/drugs/nda-and-bla-approvals/accelerated-approval-program.

`

CHAPTER 4

Mr. "BTK"

Martin Dreyling. "Mantle Cell Lymphoma: Biology, Clinical Presentation, and
Therapeutic Approaches." *American Society of Clinical Oncology Educational
Books* 4 (2014): 191–98.
Michelle Furtado, Michael L. Wang, and Brian Munneke, et al. "Ibrutinib-Associated
Lymphocytosis Corresponds to Bone Marrow Involvement in Mantle Cell
Lymphoma." *British Journal of Haematology* 170, no. 1 (2015): 131–34.
Simon Rule. Author interview, 2023.
Michael Wang. Author interview, 2013.

CHAPTER 5

"The Second Coming"

Luhua Wang, Peter Martin, and Kristie A. Blum, et al. "The Bruton's Tyrosine Kinase
Inhibitor Pci-32765 Is Highly Active as Single-Agent Therapy in Previously-Treated
Mantle Cell Lymphoma (MCL): Preliminary Results of a Phase II Trial." *Blood*
118, no. 21 (2011): 442–42.
Michael L. Wang, Simon Rule, and Peter Martin, et al. "Targeting BTK with Ibrutinib
in Relapsed or Refractory Mantle-Cell Lymphoma." *New England Journal of
Medicine* 369, no. 6 (2013): 507–16.

Window of Opportunity

M. L. Wang, P. Jain, and S. Zhao, et al. "Ibrutinib-Rituximab Followed by
R-HCVAD as Frontline Treatment for Young Patients (≤65 Years) with Mantle Cell
Lymphoma (Window-1): A Single-Arm, Phase 2 Trial." *Lancet Oncology* 23, no.
3 (2022): 406–15.

Run Happy

Maria Badillo. Author interview, 2023.
Patient Story. "Patient Stories." https://thepatientstory.com/.

The Competition: Battle of the "BTKs"

R. A. De Claro, K. M. Mcginn, and N. Verdun, et al. "FDA Approval: Ibrutinib for Patients with Previously Treated Mantle Cell Lymphoma and Previously Treated Chronic Lymphocytic Leukemia." *Clinical Cancer Research* 21, no. 16 (2015): 3586–90.

Silas Inman. "FDA Approves Ibrutinib for Chronic Lymphocytic Leukemia." *OncLive*, 2014. www.onclive.com/view/fda-approves-ibrutinib-for-cll.

Silas Inman. "FDA Approves Ibrutinib for Mantle Cell Lymphoma." *OncLive*, 2013. www.onclive.com/view/fda-approves-breakthrough-ibrutinib-for-mcl.

Going Frontline

Alice Goodman. "Ibrutinib Added to Standard Therapy Prolongs Progression-Free Survival in Older Patients with Mantle Cell Lymphoma." *ASCO Post*, 2022. https://ascopost.com/issues/june-10-2022/ibrutinib-added-to-standard-therapy-prolongs-progression-free-survival-in-older-patients-with-mantle-cell-lymphoma/#:~:text=Primary%20results%20from%20the%20phase,presentation%20at%20the%202022%20ASCO.

Johnson & Johnson Innovative Medicine. "Janssen Withdraws Application in European Union Seeking Approval of Imbruvica (Ibrutinib) for the Treatment of Patients with Untreated Mantle Cell Lymphoma," 2022. www.janssen.com/janssen-withdraws-application-european-union-seeking-approval-imbruvica-ibrutinib-treatment-patients.

N. Timofeeva and V. Gandhi. "Ibrutinib Combinations in CLL Therapy: Scientific Rationale and Clinical Results." *Blood Cancer Journal* 11, no. 4 (2021): 79.

Michael L. Wang, Wojciech Jurczak, and Mats Jerkeman, et al. "Ibrutinib Plus Bendamustine and Rituximab in Untreated Mantle-Cell Lymphoma." *New England Journal of Medicine* 386, no. 26 (2022): 2482–94.

CHAPTER 6

To Olde Towne

University of Texas MD Anderson Cancer Center. "Moon Shots Program," 2023. www.mdanderson.org/cancermoonshots.html.

White House, Office of the Press Secretary. "Cancer Moonshot," 2016.

CHAPTER 7

The Birth of a Vaccine

D. Baxby. "Edward Jenner's Inquiry after 200 Years." *BMJ* 318, no. 7180 (1999): 390.

D. Baxby. "The Jenner Bicentenary: The Introduction and Early Distribution of Smallpox Vaccine." *FEMS Immunology and Medical Microbiology* 16, no. 1 (1996): 1–10.

George Dock. "The Works of Edward Jenner and Their Value in the Modern Study of Small Pox." *New York Medical Journal*, 1902.

L. F. Haas. "Emil Adolph Von Behring (1854–1917) and Shibasaburo Kitasato (1852–1931)." *Journal of Neurology, Neurosurgery and Psychiatry* 71, no. 1 (2001): 62–62.

Edward Jenner. *An Inquiry into the Causes and Effects of the Variolae Vaccinae: A Disease Discovered in Some of the Western Counties of England, Particularly Gloucestershire, and Known by the Name of the Cow Pox.* Springfield, MA: Ashley & Brewer, 1802.

S. S. Kantha. "A Centennial Review: The 1890 Tetanus Antitoxin Paper of Von Behring and Kitasato and Related Developments." *Keio Journal of Medicine* 40, no. 1 (1991): 35–39.

Alfredo Morabia. "Edward Jenner's 1798 Report of Challenge Experiments Demonstrating the Protective Effects of Cowpox against Smallpox." *Journal of the Royal Society of Medicine* 111, no. 7 (2018): 255–57.

J. M. O'Hara, A. Yermakova, and N. J. Mantis. "Immunity to Ricin: Fundamental Insights into Toxin-Antibody Interactions." *Current Topics in Microbiology and Immunology* 357 (2012): 209–41.

John Parascandola. "The Theoretical Basis of Paul Ehrlich's Chemotherapy." *Journal of the History of Medicine and Allied Sciences* 36, no. 1 (1981): 19–43.

L. Polito, M. Bortolotti, and M. G. Battelli, et al. "Ricin: An Ancient Story for a Timeless Plant Toxin." *Toxins* 11, no. 6 (2019).

S. Riedel. "Edward Jenner and the History of Smallpox and Vaccination." *Archive of Proceedings (Baylor University Medical Center)* 18, no. 1 (2005): 21–25.

M. A. Strassburg. "The Global Eradication of Smallpox." *American Journal of Infection Control* 10, no. 2 (1982): 53–59.

Peter F. Zipfel and Christine Skerka. "From Magic Bullets to Modern Therapeutics: Paul Ehrlich, the German Immunobiologist and Physician, Coined the Term 'Complement.'" *Molecular Immunology* 150 (2022): 90–98.

The Boy from Bahia Bay

Lara Marks. "A Healthcare Revolution in the Making: The Story of César Milstein and Monoclonal Antibodies," 2023. www.whatisbiotechnology.org/index.php /exhibitions/milstein/introduction/Introduction-to-the-story-of-Cesar-Milstein-and -mAbs.

Ciser Milstein. "From the Structure of Antibodies to the Diversification of the Immune Response." *Lecture for the Nobel Prize for Physiology or Medicine*, 1984.

The Monotonous Protein

K. Horibata and A. W. Harris. "Mouse Myelomas and Lymphomas in Culture." *Experimental Cell Research* 60, no. 1 (1970): 61–77.

M. E. Jobin, J. L. Fahey, and Z. Price. "Long-Term Establishment of a Human Plasmacyte Cell Line Derived from a Patient with IGD Multiple Myeloma. I. Requirement of a Plasmacyte-Stimulating Factor for the Proliferation of Myeloma Cells in Tissue Culture." *Journal of Experimental Medicine* 140, no. 2 (1974): 494–507.

H. G. Kunkel, R. J. Slater, and R. A. Good. "Relation between Certain Myeloma Proteins and Normal Gamma Globulin." *Proceedings of the Society for Experimental Biology and Medicine* 76, no. 1 (1951): 190–93.

H. G. Kunkel and A. Tiselius. "Electrophoresis of Proteins on Filter Paper." *Journal of General Physiology* 35, no. 1 (1951): 89–118.

M. Mannik and H. G. Kunkel. "Classification of Myeloma Proteins, Bence Jones Proteins, and Macroglobulins into Two Groups on the Basis of Common Antigenic Characters." *Journal of Experimental Medicine* 116, no. 6 (1962): 859–77.

Michael Potter. "Experimental Plasmacytomagenesis in Mice." *Hematology/Oncology Clinics of North America* 11, no. 2 (1997): 323–47.

Michael Potter and Charlotte Robertson Boyce. "Induction of Plasma-Cell Neoplasms in Strain BALB/C Mice with Mineral Oil and Mineral Oil Adjuvants." *Nature* 193, no. 4820 (1962): 1086–87.

K. Rajewsky. "The Advent and Rise of Monoclonal Antibodies." *Nature* 575, no. 7781 (2019): 47–49.

Fusing One Cell Into Another

G. Köhler and C. Milstein. "Continuous Cultures of Fused Cells Secreting Antibody of Predefined Specificity." *Nature* 256, no. 5517 (1975): 495–97.

When Science Meets Commerce

Alberto Cambrosio and Peter Keating. "Between Fact and Technique: The Beginnings of Hybridoma Technology." *Journal of the History of Biology* 25, no. 2 (1992): 175–230.

A. S. Kesselheim, M. S. Sinha, and J. Avorn. "Determinants of Market Exclusivity for Prescription Drugs in the United States." *JAMA Internal Medicine* 177, no. 11 (2017): 1658–64.

B. B. Knowles. "The Wistar Legacy to Embryonic Stem Cell Research." *Monoclonal Antibodies in Immunodiagnosis and Immunotherapy* 33, no. 3 (2014): 154–57.

Lara V. Marks. "Hesitant Start: Patents, Politics, and Process." In *The Lock and Key of Medicine: Monoclonal Antibodies and the Transformation of Healthcare.* Yale University Press, 2015.

Meagen Parrish. "How Steep Is Pharma's Patent Cliff?" PharmaVoice. www .pharmavoice.com/news/pharma-patent-cliff-Merck-Keytruda-Pfizer-Seagen -Humira/652914/.

Big Biologics

Ruei-Min Lu, Yu-Chyi Hwang, and I. Ju Liu, et al. "Development of Therapeutic Antibodies for the Treatment of Diseases." *Journal of Biomedical Science* 27, no. 1 (2020): 1.

Gelu Sulugiuc. "Genmab Starts Phase I/II Humax-CD38 Trial for Myeloma." Reuters. www.reuters.com/article/genmab-myeloma/genmab-starts-phase-i-ii-humax-cd38 -trial-for-myeloma-idUSL0768049620071207.

CHAPTER 8

It Is All About the Space

T. Jelínek, J. Mihályová, and R. Hájek. "CD38 targeted Treatment for Multiple Myeloma." *Vnitr Lek* 64, no. 10 (2018): 939–48.

S. Kassem, B. K. Diallo, and N. El-Murr, et al. "SAR442085, a Novel Anti-CD38 Antibody with Enhanced Antitumor Activity against Multiple Myeloma." *Blood* 139, no. 8 (2022): 1160–76.

Henk M. Lokhorst, Torben Plesner, and Jacob P. Laubach, et al. "Targeting CD38 with Daratumumab Monotherapy in Multiple Myeloma." *New England Journal of Medicine* 373, no. 13 (2015): 1207–19.

A. Palumbo and K. Anderson. "Multiple Myeloma." *New England Journal of Medicine* 364, no. 11 (2011): 1046–60.

Niels W. C. J. Van De Donk and Saad Z. Usmani. "CD38 Antibodies in Multiple Myeloma: Mechanisms of Action and Modes of Resistance." *Frontiers in Immunology* 9 (2018).

When Myeloma Strikes

Patient Story. "Patient Stories." https://thepatientstory.com/.

The Age of Triplets and Quadruplets

B. G. M. Durie, A. Hoering, and M. H. Abidi, et al. "Bortezomib with Lenalidomide and Dexamethasone versus Lenalidomide and Dexamethasone Alone in Patients with Newly Diagnosed Myeloma without Intent for Immediate Autologous Stem-Cell Transplant (Swog S0777): A Randomised, Open-Label, Phase 3 Trial." *Lancet* 389, no. 10068 (2017): 519–27.

Henk M. Lokhorst, Torben Plesner, and Jacob P. Laubach, et al. "Targeting CD38 with Daratumumab Monotherapy in Multiple Myeloma." *New England Journal of Medicine* 373, no. 13 (2015): 1207–19.

Sagar Lonial, Brendan M. Weiss, and Saad Zafar Usmani, et al. "Daratumumab Monotherapy in Patients with Treatment-Refractory Multiple Myeloma (Sirius): An Open-Label, Randomised, Phase 2 Trial." *Lancet* 387, no. 10027 (2016): 1551–60.

Sagar Lonial, Brendan M. Weiss, and Saad Zafar Usmani, et al. "Phase II Study of Daratumumab (Dara) Monotherapy in Patients with ≥ 3 Lines of Prior Therapy or Double Refractory Multiple Myeloma (MM): 54767414mmy2002 (Sirius)." *Journal of Clinical Oncology* 33, no. 18 (2015): LBA8512–LBA8512.

H. Ludwig, S. Novis Durie, and A. Meckl, et al. "Multiple Myeloma Incidence and Mortality around the Globe; Interrelations between Health Access and Quality, Economic Resources, and Patient Empowerment." *Oncology* 25: e1406–e1413.

S. V. Rajkumar. "Doublets, Triplets, or Quadruplets of Novel Agents in Newly Diagnosed Myeloma?" *Hematology American Society of Hematology Education Program* (2012): 354–61.

Targeted Oncology. "Treatment Rationale for Using Triplets vs Quadruplets in Multiple Myeloma," 2023. www.targetedonc.com/view/treatment-rationale-for-using-triplets-vs-quadruplets-in-multiple-myeloma.

S. Z. Usmani, B. M. Weiss, and T. Plesner, et al. "Clinical Efficacy of Daratumumab Monotherapy in Patients with Heavily Pretreated Relapsed or Refractory Multiple Myeloma." *Blood* 128, no. 1 (2016): 37–44.

CHAPTER 9

It Takes One Patient

T. M. Herndon, A. Deisseroth, and E. Kaminskas, et al. "U.S. Food & Drug Administration Approval: Carfilzomib for the Treatment of Multiple Myeloma." *Clinical Cancer Research* 19, no. 17 (2013): 4559–63.

A. Hoter, M. E. El-Sabban, and H. Y. Naim. "The HSP90 Family: Structure, Regulation, Function, and Implications in Health and Disease." *International Journal of Molecular Science* 19, no. 9 (2018).

C. Hu, J. Yang, and Z. Qi, et al. "Heat Shock Proteins: Biological Functions, Pathological Roles, and Therapeutic Opportunities." *MedComm* 3, no. 3 (2022): e161.

Saad Z. Usmani. Author interview, 2023.

Saad Z. Usmani, Imran Khan, and Christopher Chiu, et al. "Deep Sustained Response to Daratumumab Monotherapy Associated with T-Cell Expansion in Triple Refractory Myeloma." *Experimental Hematology & Oncology* 7, no. 1 (2018): 3.

Enemy Unseen

Deborah Delaune, Vibol Hul, and Erik A. Karlsson, et al. "A Novel SARS-CoV-2 Related Coronavirus in Bats from Cambodia." *Nature Communications* 12, no. 1 (2021): 6563.
Paul G. Richardson, Michel Attal, and S. Vincent Rajkumar, et al. "A Phase III Randomized, Open Label, Multicenter Study Comparing Isatuximab, Pomalidomide, and Low-Dose Dexamethasone versus Pomalidomide and Low-Dose Dexamethasone in Patients with Relapsed/Refractory Multiple Myeloma (RRMM)." *Journal of Clinical Oncology* 37, no. 15 (2019): 8004.
G. Salles, A. K. Gopal, and M. C. Minnema, et al. "Phase 2 Study of Daratumumab in Relapsed/Refractory Mantle-Cell Lymphoma, Diffuse Large B-Cell Lymphoma, and Follicular Lymphoma." *Clinical Lymphoma Myeloma Leukemia* 19, no. 5 (2019): 275–84.
Hengbo Zhu, Li Wei, and Ping Niu. "The Novel Coronavirus Outbreak in Wuhan, China." *Global Health Research and Policy* 5, no. 1 (2020): 6.

The Platelet Conundrum

D. J. Angiolillo, M. Ueno, and S. Goto. "Basic Principles of Platelet Biology and Clinical Implications." *Circulation* 74, no. 4 (2010): 597–607.
Paola E. J. Van Der Meijden and Johan W. M. Heemskerk. "Platelet Biology and Functions: New Concepts and Clinical Perspectives." *Nature Reviews Cardiology* 16, no. 3 (2019): 166–79.

CHAPTER 10

First Approval

Lee Greenburger. "FDA Approves New Therapy for Patients with Previously Treated Multiple Myeloma." Leukemia & Lymphoma Society, 2020. https://tlls .org/blog/fda-approves-new-therapy-for-patients-with-previously-treated-multiple -myeloma.
U.S. Food & Drug Administration. "FDA Approves Isatuximab-Irfc for Multiple Myeloma," 2020. www.fda.gov/drugs/resources-information-approved-drugs/fda -approves-isatuximab-irfc-multiple-myeloma-0#:~:text=On%20March%202%2C %202020%2C%20the,lenalidomide%20and%20a%20proteasome%20inhibitor.

A Trial in Solitude

CBS. "Italy Imposes Unprecedented Lockdown to Fight Coronavirus: 'We Have Run out of Time.'" *CBS Mornings*, 2020. www.cbsnews.com/news/coronavirus-italy -quarantine-lockdown-prime-minister-we-have-run-out-of-time/.

Danilo Cereda, Mattia Manica, and Marcello Tirani, et al. "The Early Phase of the Covid-19 Epidemic in Lombardy, Italy." *Epidemics* 37 (2021).

Rachel Donadio. "I Can't Stop Thinking about Patient One." *Atlantic*, 2020. www.theatlantic.com/international/archive/2020/04/italy-patient-one-family-coronavirus-covid19/610039/.

A. Maruotti, G. Jona-Lasinio, and F. Divino, et al. "Estimating Covid-19-Induced Excess Mortality in Lombardy, Italy." *Aging Clinical Experimental Research* 34, no. 2 (2022): 475–79.

Norimitsu Onishi. "Chaos in Europe, and Anger, Over U.S. Travel Ban to Curb Coronavirus." *New York Times*, 2020. www.nytimes.com/2020/03/12/world/europe/europe-coronavirus-travel-ban.html.

"It's Apocalyptic"

Kaitlyn Folmer and Aaron Katersky. "Inside the Coronavirus 'Ground Zero': Elmhurst Hospital in New York City." *ABC News*, 2020. https://abcnews.go.com/Politics/inside-coronavirus-ground-elmhurst-hospital-york-city/story?id=69804681.

Y. Huang and R. Li. "The Lockdown, Mobility, and Spatial Health Disparities in Covid-19 Pandemic: A Case Study of New York City." *Cities* 122 (2022).

KFF Health News. "First Moderna Shots Given: Biontech Aims to Ramp up 2021 Production," 2020. https://kffhealthnews.org/morning-breakout/first-moderna-shots-given-biontech-aims-to-ramp-up-2021-production/.

Peter Loftus and Melanie Grayce West. "First Covid-19 Vaccine Given to U.S. Public." *Wall Street Journal*, 2020.

Hyuna Sung, Jacques Ferlay, and Rebecca L. Siegel, et al. "Global Cancer Statistics 2020: Globocan Estimates of Incidence and Mortality Worldwide for 36 Cancers in 185 Countries." *CA* 71, no. 3 (2021): 209–49.

World Health Organization. "The True Death Toll of Covid-19: Estimating Global Excess Mortality," 2020. www.who.int/data/stories/the-true-death-toll-of-covid-19-estimating-global-excess-mortality.

CHAPTER 11

Cancer Conspiracies

Ying-Yi Chen, Kuan-Hsun Lin, and Yen-Shou Kuo, et al. "Therapeutic Impact of Epidermal Growth Factor Receptor Tyrosine Kinase Inhibitor with Various Treatment Combinations for Advanced Lung Adenocarcinoma." *World Journal of Surgical Oncology* 21, no. 1 (2023): 326.

Peter J. O'Dwyer, Robert J. Gray, and Keith T. Flaherty, et al. "The NCI-Match Trial: Lessons for Precision Oncology." *Nature Medicine* 29, no. 6 (2023): 1349–57.

L. K. Ursell, J. L. Metcalf, L. W. Parfrey, and R. Knight. "Defining the Human Microbiome." *Nutrition Review* 70, supp. 1 (2012): S38–S44.

How Cancer Escapes

Niels W. C. J. Van De Donk and S. Z. Usmani. "CD38 Antibodies in Multiple Myeloma: Mechanisms of Action and Modes of Resistance." *Frontiers in Immunology* 9 (2018): 2134.

Neil Vasan, José Baselga, and David M. Hyman. "A View on Drug Resistance in Cancer." *Nature* 575, no. 7782 (2019): 299–309.

Hasta la Vista

Kyle LaHucik. "Sanofi Ends 2 Phase 1 Cancer Assets Weeks after Axing Sangamo Deal," 2022. www.fiercebiotech.com/biotech/sanofi-ends-two-phase-1-cancer -assets-weeks-after-axing-sangamo-deal-prioritizing-covid-19.

"Tumor-Meter"

Greg Brozeit. "IMW: Challenges Remain Despite Progress in Myeloma." HealthTree Foundation for Multiple Myeloma, 2021. https://healthtree.org/myeloma /community/articles/challenges-remain-despite-progress-in-myeloma.

E. Giovannucci, D. M. Harlan, and M. C. Archer, et al. "Diabetes and Cancer: A Consensus Report." *Diabetes Care* 33, no. 7 (2010): 1674–85.

Y. Hwangbo, D. Kang, and M. Kang, et al. "Incidence of Diabetes after Cancer Development: A Korean National Cohort Study." *JAMA Oncology* 4, no. 8 (2018): 1099–1105.

S. A. Leon, B. Shapiro, D. M. Sklaroff, and M. J. Yaros. "Free DNA in the Serum of Cancer Patients and the Effect of Therapy." *Cancer Research* 37, no. 3 (1977): 646–50.

Christopher Weaver. "Agony, Alarm and Anger for People Hurt by Theranos's Botched Blood Tests." *Wall Street Journal*, 2016.

B. Zhu and S. Qu. "The Relationship between Diabetes Mellitus and Cancers and Its Underlying Mechanisms." *Frontiers in Endocrinology* 13 (2022): 800–995.

CHAPTER 12

From the Ashes

Jason Dawsey. "A Shared Enmity: Germany, Japan, and the Creation of the Tripartite Pact." National World War II Museum. www.nationalww2museum.org/war/articles /germany-japan-tripartite-pact.

Ingemar Ernberg, Klas Kärre, and Hans Wigzell. "George Klein (1925–2016)." *Nature* 542, no. 7641 (2017): 296–96.

Kafkadesk Budapest. *On This Day, in 1944: Nazi Germany Invaded Hungary*, in *Kafkadesk*, 2021.

George Klein and Eva Klein. "How One Thing Has Led to Another." *Annual Review of Immunology* 7, no. 1 (1989): 1–34.
Leonard Norkin. *Virology: Molecular Biology and Pathogenesis*. Washington, DC: ASM Press, 2010.
Sasvári Péter. "'I Did Not Hold You Back, I Am Guilty': The Life and Tragedy of Pál Teleki." *Hungarian Conservative*, 2023.
Pramod K. Srivastava. "George Klein (1925–2016) a Prescient, Luminous Voice." *Cancer Immunology Research* 5, no. 4 (2017): 272–72.

The Elusive Immunity

Paul Ehrlich. "Croonian Lecture: On Immunity with Special Reference to Cell Life." *Proceedings of the Royal Society of London* 66 (1900): 424–48.
Paul Ehrlich. "Experimentelle Studien an Mäusetumoren." *Zeitschrift für krebsforschung* 5, no. 2 (1907): 59–81.
G. Klein, H. O. Sjogren, E. Klein, and K. E. Hellstrom. "Demonstration of Resistance against Methylcholanthrene-Induced Sarcomas in the Primary Autochthonous Host." *Cancer Research* 20 (1960): 1561–72.

"A Useless Organ"

D. H. D. Gray, D. L. Vaux, and A. Strasser. "The 2019 Lasker Award: T Cells and B Cells, Whose Life and Death Are Essential for Function of the Immune System." *Cell Death & Differentiation* 26, no. 12 (2019): 2513–15.
Jacques Miller. "The Early Work on the Discovery of the Function of the Thymus: An Interview with Jacques Miller." *Cell Death & Differentiation* 27, no. 1 (2020): 396–401.
J. F. Miller. "Recovery of Leukaemogenic Agent from Nonleukaemic Tissues of Thymectomized Mice." *Nature* 187 (1960): 703.
J. F. Miller. "Studies on Mouse Leukaemia." The Fate of Thymus Homografts in Immunologically Tolerant Mice. *British Journal of Cancer* 14, no. 2 (1960): 244–55.
J. F. Miller, G. A. Grant, and F. J. Roe. "Effect of Thymectomy on the Induction of Skin Tumours by 3,4-Benzopyrene." *Nature* 199 (1963): 920–22.
J. F. Miller and A. P. Miller. " Role of the Thymus in Murine Leukæmia." *Nature* 183, no. 4667 (1959): 1069.

CHAPTER 13

(Reverse) Namesake

Neil Canavan. *A Cure Within: Scientists Unleashing the Immune System to Kill Cancer*. Cold Spring Harbor, NY: Cold Spring Harbor Laboratory Press, 2017.

Mark M. Davis, Yueh-Hsiu Chien, Nicholas R. J. Gascoigne, and Stephen M. Hedrick. "A Murine T Cell Receptor Gene Complex: Isolation, Structure and Rearrangement." *Immunological Reviews* 81, no. 1 (1984): 235–58.

Tak W. Mak. "From the T-Cell Receptor to Cancer Therapy: An Interview with Tak W." Mak. *Cell Death & Differentiation* 28, no. 1 (2021): 5–14.

Tak Mak. "The Future of Cancer Treatment as Seen by Tak Mak, Whose Lab Paved the Way for T- Cell Receptor Based Immunotherapies." *Cancerworld*, 2023.

T. W. Mak. "The T-Cell Antigen Receptor: 'The Hunting of the Snark.'" *European Journal of Immunology* 37, supp. 1 (2007): S83–S93.

Peter Valent, Bernd Groner, and Udo Schumacher, et al. "Paul Ehrlich (1854–1915) and His Contributions to the Foundation and Birth of Translational Medicine." *Journal of Innate Immunity* 8, no. 2 (2016): 111–20.

Yusuke Yanagi, Yasunobu Yoshikai, and Kathleen Leggett, et al. "A Human T Cell–Specific CDNA Clone Encodes a Protein Having Extensive Homology to Immunoglobulin Chains." *Nature* 308, no. 5955 (1984): 145–49.

Self Recognition

Blau Ca. "E. Donnall Thomas, M.D. (1920–2012)." *Stem Cells Translational Medicine* 2, no. 2 (2013): 81–82.

C. H. June. "Toward Synthetic Biology with Engineered T Cells: A Long Journey Just Begun." *Human Gene Therapy* 25, no. 9 (2014): 779–84.

Carl H. June, Roddy S. O'Connor, and Omkar U. Kawalekar, et al. "CAR T Cell Immunotherapy for Human Cancer." *Science* 359, no. 6382 (2018): 1361–65.

Viruses as Tools

Phillip Boffey. "Reagan Defends Financing for Aids." *New York Times*, 1985.

Jason M. Brenchley, Brenna J. Hill, and David R. Ambrozak, et al. "T-Cell Subsets That Harbor Human Immunodeficiency Virus (HIV) in Vivo: Implications for HIV Pathogenesis." *Journal of Virology* 78, no. 3 (2004): 1160–68.

Mary Engel. "Dr. Carl June Weaves Together HIV and Cancer Research to Advance Cures for Both." Fred Hutch Cancer Center. www.fredhutch.org/en/news/center -news/2017/08/carl-june-weaves-together-hiv-and-cancer-research-to-advance -cures-for-both.html.

Carl H. June and Michel Sadelain. "Chimeric Antigen Receptor Therapy." *New England Journal of Medicine* 379, no. 1 (2018): 64–73.

Robert Pear. "Health Chief Calls Aids Battle 'No. 1 Priority.'" *New York Time*, 1983.

Maup van de Kerkhof. "The Chimera: The Greek Monster Challenging the Imaginable." History Cooperative. https://historycooperative.org/the-chimera/.

Before CAR

Steven G. Deeks, Bridget Wagner, and Peter A. Anton, et al. "A Phase II Randomized Study of HIV-Specific T-Cell Gene Therapy in Subjects with Undetectable Plasma Viremia on Combination Antiretroviral Therapy." *Molecular Therapy* 5, no. 6 (2002): 788–97.

Zelig Eshhar, Tova Waks, Gideon Gross, and Daniel G Schindler. "Specific Activation and Targeting of Cytotoxic Lymphocytes through Chimeric Single Chains Consisting of Antibody-Binding Domains and the Gamma or Zeta Subunits of the Immunoglobulin and T-Cell Receptors." *Proceedings of the National Academy of Sciences* 90, no. 2 (1993): 720–24.

G. Gross, T. Waks, and Z. Eshhar. "Expression of Immunoglobulin-T-Cell Receptor Chimeric Molecules as Functional Receptors with Antibody-Type Specificity." *Proceedings of the National Academy of Sciences* 86, no. 24 (1989): 10024–28.

Ronald T. Mitsuyasu, Peter A. Anton, and Steven G. Deeks, et al. "Prolonged Survival and Tissue Trafficking Following Adoptive Transfer of Cd4ζ Gene-Modified Autologous Cd4+ and Cd8+ T Cells in Human Immunodeficiency Virus–Infected Subjects." *Blood* 96, no. 3 (2000): 785–93.

D. A. Morgan, F. W. Ruscetti, and R. Gallo. "Selective in Vitro Growth of T Lymphocytes from Normal Human Bone Marrows." *Science* 193, no. 4257 (1976): 1007–8.

Steven A. Rosenberg. "A Journey in Science: Immersion in the Search for Effective Cancer Immunotherapies." *Molecular Medicine* 27, no. 1 (2021): 63.

Steven A. Rosenberg, Michael T. Lotze, and Linda M Muul, et al. "Observations on the Systemic Administration of Autologous Lymphokine-Activated Killer Cells and Recombinant Interleukin-2 to Patients with Metastatic Cancer." *New England Journal of Medicine* 313, no. 23 (1985): 1485–92.

Steven A. Rosenberg, Paul Spiess, and Rene Lafreniere. "A New Approach to the Adoptive Immunotherapy of Cancer with Tumor-Infiltrating Lymphocytes." *Science* 233, no. 4770 (1986): 1318–21.

Robert E. Walker, Christine M Bechtel, and Ven Natarajan, et al. "Long-Term in Vivo Survival of Receptor-Modified Syngeneic T Cells in Patients with Human Immunodeficiency Virus Infection." *Blood* 96, no. 2 (2000): 467–74.

Wall Street Journal. "Cell Genesys Shares Surge on Outlook for Aids Therapy," 1997.

CHAPTER 14

A Gas Pedal for CAR

Michael H. Kershaw, Jennifer A. Westwood, and Linda L. Parker, et al. "A Phase I Study on Adoptive Immunotherapy Using Gene-Modified T Cells for Ovarian Cancer." *Clinical Cancer Research* 12, no. 20 (2006): 6106–15.

Cor H. J. Lamers, Stefan Sleijfer, and Arnold G. Vulto, et al. "Treatment of Metastatic Renal Cell Carcinoma with Autologous T-Lymphocytes Genetically Retargeted against Carbonic Anhydrase IX: First Clinical Experience." *Journal of Clinical Oncology* 24, no. 13 (2006): e20–e22.

Charlotte Schubert. "Juno's Lasting Legacy: How the Cell Therapy Juggernaut Influenced Biotech in Seattle and Beyond." GeekWire. www.geekwire.com /2022/junos-lasting-legacy-how-the-cell-therapy-juggernaut-influenced-biotech-in -seattle-and-beyond/.

A Three-Way Race

Vicki Brower. "The CAR T-Cell Race." *Scientist*, 2015.

James N. Kochenderfer, Wyndham H. Wilson, and John E. Janik, et al. "Eradication of B-Lineage Cells and Regression of Lymphoma in a Patient Treated with Autologous T Cells Genetically Engineered to Recognize CD19." *Blood* 116, no. 20 (2010): 4099–4102.

James N. Kochenderfer, Zhiya Yu, and Dorina Frasheri, et al. "Adoptive Transfer of Syngeneic T Cells Transduced with a Chimeric Antigen Receptor That Recognizes Murine CD19 Can Eradicate Lymphoma and Normal B Cells." *Blood* 116, no. 19 (2010): 3875–86.

Doug Olson. "A Breakthrough in Cancer Treatment—A Patient's Story." CLL Society, 2018. https://cllsociety.org/2018/06/a-breakthrough-in-cancer-treatment-a -patients-story/.

Penn Medicine News. "Study of Penn Patients with Decade-Long Leukemia Remissions after CAR T Cell Therapy Reveals New Details about Persistence of Personalized 'Living Drug.'" www.pennmedicine.org/news/news-releases/2022 /february/study-of-penn-patients-with-decade-long-leukemia-remissions-after-car -t-cell-therapy.

Andrew Pollack. "Setting the Body's 'Serial Killers' Loose on Cancer." *New York Times*, 2016. www.nytimes.com/2016/08/02/health/cancer-cell-therapy-immune -system.html.

D. L. Porter, B. L. Levine, and M. Kalos, et al. "Chimeric Antigen Receptor-Modified T Cells in Chronic Lymphoid Leukemia." *New England Journal of Medicine* 365, no. 8 (2011): 725–33.

Antonio Regalado. "T-Cell Pioneer Carl June Acknowledges Key Ingredient Wasn't His." MIT Technology Review. www.technologyreview.com/2016/03/14/161592/t -cell-pioneer-carl-june-acknowledges-key-ingredient-wasnt-his/.

CHAPTER 15

The First Child

Emily Whitehead Foundation. "Our Journey." https://emilywhiteheadfoundation.org /our-journey/.

Ashley Moore. "Emily Whitehead, First Pediatric Patient to Receive CAR T-Cell Therapy, Celebrates Cure 10 Years Later." Children's Hospital of Philadelphia,

Pennsylvania, www.chop.edu/news/emily-whitehead-first-pediatric-patient
-receive-car-t-cell-therapy-celebrates-cure-10-years.

A. Neaga, L. Jimbu, and O. Mesaros, et al. "Why Do Children with Acute
Lymphoblastic Leukemia Fare Better Than Adults? *Cancers* 13, no. 15 (2021).

Jessica Skarzynski. "Faith, Science, and Miracles: A Conversation about CAR–T
Cell Therapy with the Whitehead Family." *Cure*. www.curetoday.com/view/faith
-science-and-miracles-a-conversation-about-car-t-cell-therapy-with-the-whitehead
-family.

T. Terwilliger and M. Abdul-Hay. "Acute Lymphoblastic Leukemia: A Comprehensive
Review and 2017 Update." *Blood Cancer Journal* 7, no. 6 (2017): e577.

Kari Whitehead, Tom Whitehead, and Emily Whitehead. *Praying for Emily: The
Faith, Science, and Miracles That Saved Our Daughter*. New York: Worthy
Books, 2020.

A Cytokine Storm

Children's Hospital of Philadelphia Research Institute. "Emily Whitehead, 10 Years
Later: Q&A with Stephan Grupp, MD, PhD." www.research.chop.edu/cornerstone
-blog/emily-whitehead-10-years-later-qa-with-stephan-grupp-md-phd.

Denise Grady. "In Girl's Last Hope, Altered Immune Cells Beat Leukemia." *New
York Times*, 2012. www.nytimes.com/2012/12/10/health/a-breakthrough-against
-leukemia-using-altered-t-cells.html.

Daniel W. Lee, Rebecca Gardner, and David L. Porter, et al. "Current Concepts in
the Diagnosis and Management of Cytokine Release Syndrome." *Blood* 124, no.
2 (2014): 188–95.

Tom Whitehead. "Against All Odds." *ASCO Post*. https://ascopost.com/issues
/january-25-2018/against-all-odds/.

CARs on the Move

M. L. Davila, I. Riviere, and X. Wang, et al. "Efficacy and Toxicity Management of
19–28Z CAR T Cell Therapy in B Cell Acute Lymphoblastic Leukemia." *Science
Translational Medicine* 6, no. 224 (2014).

D. W. Lee, J. N. Kochenderfer, and M. Stetler-Stevenson, et al. "T Cells Expressing
CD19 Chimeric Antigen Receptors for Acute Lymphoblastic Leukaemia in
Children and Young Adults: A Phase 1 Dose-Escalation Trial." *Lancet* 385, no.
9967 (2015): 517–28.

Shannon L. Maude, Noelle Frey, and Pamela A. Shaw, et al. "Chimeric Antigen
Receptor T Cells for Sustained Remissions in Leukemia." *New England Journal of
Medicine* 371, no. 16 (2014): 1507–17.

S. L. Maude, D. T. Teachey, D. L. Porter, and S. A. Grupp. "CD19–Targeted Chimeric
Antigen Receptor T-Cell Therapy for Acute Lymphoblastic Leukemia." *Blood* 125,
no. 26 (2015): 4017–23.

Sattva Neelapu. Author interview, 2023.

Race to the Finish Line

John Carroll. "Kite Pharma Doubles Down on Crucial NCI Partnership, Expands Work on CAR–T/TCRS." Fierce Biotech. www.fiercebiotech.com/partnering/kite-pharma-doubles-down-on-crucial-nci-partnership-expands-work-on-car-t-tcrs.

Gilead. "Kite Pharma Announces Exclusive License with the National Institutes of Health for Fully Human Anti-CD19 Chimeric Antigen Receptor (CAR) Product Candidate to Treat B-Cell Malignancies." www.gilead.com/news-and-press/press-room/press-releases/2016/7/kite-pharma-announces-exclusive-license-with-the-national-institutes-of-health-for-fully-human-anticd19-chimeric-antigen-receptor-car-product-candi.

Emma C. Morris, Sattva S. Neelapu, Theodoros Giavridis, and Michel Sadelain. "Cytokine Release Syndrome and Associated Neurotoxicity in Cancer Immunotherapy." *Nature Reviews Immunology* 22, no. 2 (2022): 85–96.

Sattva S. Neelapu. "Managing the Toxicities of CAR T-Cell Therapy." *Hematological Oncology* 37, supp. S1 (2019): 48–52.

Sattva S. Neelapu, Frederick L. Locke, and Nancy L. Bartlett, et al. "Axicabtagene Ciloleucel Car T-Cell Therapy in Refractory Large B-Cell Lymphoma." *New England Journal of Medicine* 377, no. 26 (2017): 2531–44.

Sattva S. Neelapu, Frederick L. Locke, and Nancy L. Bartlett, et al. "KTE-C19 (Anti-CD19 Car T Cells) Induces Complete Remissions in Patients with Refractory Diffuse Large B-Cell Lymphoma (DLBCL): Results from the Pivotal Phase 2 Zuma-1." *Blood* 128, no. 22 (2016).

E. L. Siegler and S. S. Kenderian. "Neurotoxicity and Cytokine Release Syndrome after Chimeric Antigen Receptor T Cell Therapy: Insights into Mechanisms and Novel Therapies." *Frontiers in Immunology* 11 (2020): 1973.

CHAPTER 16

First Approval

Francis Collins, MD. "FDA Approves First CAR–T Cell Therapy for Pediatric Acute Lymphoblastic Leukemia." NIH Director's Blog. https://directorsblog.nih.gov/2017/08/30/fda-approves-first-car-t-cell-therapy-for-pediatric-acute-lymphoblastic-leukemia/.

Gilead. "Kite's Yescarta (axicabtagene ciloleucel) Becomes First CAR–T Therapy Approved by the FDA for the Treatment of Adult Patients with Relapsed or Refractory Large B-Cell Lymphoma after Two or More Lines of Systemic Therapy." *Busines Wire*, 2017.

S. L. Maude, T. W. Laetsch, and J. Buechner, et al. "Tisagenlecleucel in Children and Young Adults with B-Cell Lymphoblastic Leukemia." *New England Journal of Medicine* 378, no. 5 (2018): 439–48.

National Cancer Institute. "CAR–T Cells: Engineering Patients' Immune Cells to Treat Their Cancers." www.cancer.gov/about-cancer/treatment/research/car-t-cells.

Novartis. "Novartis Receives First Ever FDA Approval for a CAR–T Cell Therapy, Kymriah (CTL019), for Children and Young Adults with B-Cell ALL That Is Refractory or Has Relapsed at Least Twice." www.novartis.com/news/media -releases/novartis-receives-first-ever-fda-approval-car-t-cell-therapy-kymriahtm -ctl019-children-and-young-adults-b-cell-all-refractory-or-has-relapsed-least -twice#:~:text=Basel%2C%20August%2030%2C%202017%20%2D,of%20age %20with%20B%2Dcell.

U.S. Food & Drug Administration. "FDA Approves Tisagenlecleucel for B-Cell ALL and Tocilizumab for Cytokine Release Syndrome." www.fda.gov/drugs /resources-information-approved-drugs/fda-approves-tisagenlecleucel-b-cell-all -and-tocilizumab-cytokine-release-syndrome.

A Complicated Choice

BMT Infonet.org. "Directory of Car T-Cell Therapy Centers." www.bmtinfonet.org /car-t-cell-therapy-center-directory.

Sattva S. Neelapu, Michael Dickinson, and Javier Munoz, et al. "Axicabtagene Ciloleucel as First-Line Therapy in High-Risk Large B-Cell Lymphoma: The Phase 2 Zuma-12 Trial." *Nature Medicine* 28, no. 4 (2022): 735–42.

"Cover Me in Sunshine"

NBC. "First Child to Receive CAR T-Cell Therapy Is 10 Years Cancer-Free." *Today Show*, 2022. www.today.com/video/first-child-to-receive-car-t-cell-therapy-is-10 -years-cancer-free-139772997670.

Pink. "Cover Me in Sunshine." *All I Know So Far*, Maureen Mcdonald and Amy Allen. RCA Records, 2021.

CHAPTER 17

Achy Fingers

Patient Story. "William's B-Cell Acute Lymphoblastic Leukemia Story." https: //thepatientstory.com/patient-stories/leukemia/acute-lymphoblastic-leukemia-all /william-y/.

Immune Escape

A. C. Anderson, N. Joller, and V. K. Kuchroo. "LAG-3, TIM-3, and TIGIT: Co-Inhibitory Receptors with Specialized Functions in Immune Regulation." *Immunity* 44, no. 5 (2016): 989–1004.

Sattva S. Neelapu, John M. Rossi, and Caron A. Jacobson, et al. "CD19-Loss with Preservation of Other B Cell Lineage Features in Patients with Large B Cell Lymphoma Who Relapsed Post-Axi-Cel." *Blood* 134, supp. 1 (2019): 203.

Robert C. Sterner and Rosalie M. Sterner. "CAR–T Cell Therapy: Current Limitations and Potential Strategies." *Blood Cancer Journal* 11, no. 4 (2021): 69.

CHAPTER 18

ASH Clinical News. "CAR T-Cell Therapies Predicted to Cost More Than $1 Million Per Patient." https://ashpublications.org/ashclinicalnews/news/3469/CAR-T-Cell -Therapies-Predicted-to-Cost-More-Than-1#.

G. Choi, G. Shin, and S. Bae. "Price and Prejudice? The Value of Chimeric Antigen Receptor (CAR) T-Cell Therapy." *International Journal of Environmental Research and Public Health* 19, no. 19 (2022).

S. Depil, P. Duchateau, and S. A. Grupp, et al. "'Off-the-Shelf' Allogeneic CAR T Cells: Development and Challenges." *Nature Reviews Drug Discovery* 19, no. 3 (2020): 185–99.

Fraiser Kansteiner. "Atara Makes History with World-First Nod for Allogeneic T-Cell Therapy Ebvallo." Fierce Pharma, 2022. www.fiercepharma.com/pharma/atara -makes-history-world-first-nod-allogeneic-t-cell-therapy-ebvallo.

J. K. Lin, B. J. Lerman, and J. I. Barnes, et al. "Cost Effectiveness of Chimeric Antigen Receptor T-Cell Therapy in Relapsed or Refractory Pediatric B-Cell Acute Lymphoblastic Leukemia." *Journal of Clinical Oncology* 36, no. 32 (2018): 3192–202.

Bispecific Bandwagon

Tara Arvedson, Julie M. Bailis, Thomas Urbig, and Jennitte L. Stevens. "Considerations for Design, Manufacture, and Delivery for Effective and Safe T-Cell Engager Therapies." *Current Opinion in Biotechnology* 78 (2022).

Ameet Patel, Olalekan Oluwole, Bipin Savani, and Bhagirathbhai Dholari. "Taking a Bite out of the CAR T Space Race." *British Journal of Haematology* 195, no. 5 (2021): 689–97.

Shujie Zhou, Mingguo Liu, and Fei Ren, et al. "The Landscape of Bispecific T Cell Engager in Cancer Treatment." *Biomarker Research* 9, no. 1 (2021): 38.

CHAPTER 19

American Cancer Society. "Lifetime Risk of Developing or Dying from Cancer." www .cancer.org/cancer/risk-prevention/understanding-cancer-risk/lifetime-probability -of-developing-or-dying-from-cancer.html.

Mathew Weaver and agencies. "Cancer Survival Rates Have Doubled since 1970s, Research Shows." *Guardian*, 2010. www.theguardian.com/science/2010/jul/12 /cancer-survival-rates-doubled#:~:text=On%20average%20it%20found%20that ,20%25%20in%201971%2D72.

EPILOGUE

Ryan Cross. "This $25,000 Physical Has Found Some 'Serious' Health Problems: Others Say It Has Serious Problems." *Science*, 2017. www.science.org /content/article/25000-physical-has-found-some-serious-health-problems-others -say-it-has-serious.

Chris Crouse. *Genome Pioneer Craig Venter Is Trying to Decode Death.* CNBC, 2018. www.cnbc.com/2018/03/27/genome-pioneer-craig-venter-is-trying-to -decode-death.html#:~:text=J.,lurking%20within%20seemingly%20healthy %20individuals.

Ying-Chen Claire Hou, Hung-Chun Yu, and Rick Martin, et al. "Precision Medicine Integrating Whole-Genome Sequencing, Comprehensive Metabolomics, and Advanced Imaging." *Proceedings of the National Academy of Sciences* 117, no. 6 (2020): 3053–62.

Human Longevity. "The Human Longevity Difference." https://humanlongevity.com /the-hli-difference/.

Alex Keown. "Corvus CEO Richard Miller Wins Drug Hunter Award." BioSpace, 2020. www.biospace.com/article/corvus-ceo-richard-miller-wins-drug-hunter -award/.

Index

tetanus, 65
thalidomide, 36
Thatcher, Margaret, 73–74
Theranos, 118
Thomas, E. Donnall, 136
thrombocytes (platelets), 101–3
thrombocytopenia, 102, 103
thymus: lymphocytic leukemia of, 130–31; T-cells in, 135
TIL. *See* tumor infiltrating lymphocytes
TIM3, 168
Tisa-cel, 159, 169–70
tissue plasminogen activator (TPA), 18
TKIs. *See* tyrosine kinase inhibitors
tocilizumab, 151–52, 156–57; FDA and, 159
Tolstoy, Leo, 61
toxicity: of CAR-T, 153–56, 161, 171; of chemotherapy, 37, 46, 50, 165; DLT, 97; dose-response relationship and, 81; of mAb85, 80
toxins: antitoxins and, 65–66; B-cells and, 15, 32; of diphtheria and tetanus, 65
TPA. *See* tissue plasminogen activator
Trump, Donald, 107
tumor infiltrating lymphocytes (TIL), 139
tumor stroma, 139
tyrosine kinase inhibitors (TKIs): biomarkers for, 112; for lung cancer, 112; for lymphoma, 20
tyrosine kinases, 11; B-cell lymphoma and, 17; fostamatinib and, 18
tyrosine kinases enzyme blockers, for lymphoma, 15

Usmani, Saad, 89–94

vaccines: for COVID-19, 110; Jenner and, 61–64, 120, 129; Klein and Fisher and, 128; for smallpox, 62–64, 129
variable regions, of antibodies, 68
variolation, 62, 63

Velcade, 84
Velcade-Revlimid (VR) with dexamethasone, 86
Venter, Craig, 3, 5–7, 10–13, 39, 113; Human Longevity Initiative and, 177; ibrutinib and, 40–41
Vincerx Pharma, 178
viruses: B-cells and, 15, 32; CRS and, 151; killer T-cells and, 93; kinases and, 17; lymphocytic leukemia from, 130–31; lymphoma and, 17; of *Rhinolophus* bats, 98. *See also specific viruses*
VR. *See* Velcade-Revlimid
Vrba, Rudolf, 124

Wang, Michael, 31–36, 38, 41–42, 43, 45, 48, 49, 100, 156, 178; BTK and, 33; Hamdy and, 32–33; MCL and, 32–36; R2 and, 32; SHINE study and, 53–54
Wellcome Trust, 5
Wetzler, Alfred, 124
white blood cells (lymphocytes): daratumumab and, 93–94; dying of, 8; ibrutinib and, 37; IL2 and, 138–39; mAb85 and, 101; MCL and, 38; MM and, 69–70; PCI 32765 and, 26–27; SYK and, 20. *See also* B-cells; T-cells
Whitehead, Emily, 120, 147–52, 153, 156, 159–60, 162, 179
WHO. *See* World Health Organization
window trials, for MCL, 44, 46, 48–49
World Health Organization (WHO), 64, 100; on COVID-19, 106

Xcytrin, 21–22, 25

Yanagi, Yusuke, 134
Yescarta, 160

zanubrutinib, 51
Zauberkugeln (magic-bullet), 9, 42, 49, 64, 66